Simon Rothfuß

Human-Machine Cooperative Decision Making

Karlsruher Beiträge zur
Regelungs- und Steuerungstechnik
Karlsruher Institut für Technologie

Band 19

Human-Machine Cooperative Decision Making

by
Simon Rothfuß

Scientific Publishing

Karlsruher Institut für Technologie
Institut für Regelungs- und Steuerungssysteme

Human-Machine Cooperative Decision Making

Zur Erlangung des akademischen Grades eines Doktor-Ingenieurs von der KIT-Fakultät für Elektrotechnik und Informationstechnik des Karlsruher Instituts für Technologie (KIT) genehmigte Dissertation

von Simon Rothfuß, M.Sc.

Tag der mündlichen Prüfung: 4. März 2022
Hauptreferent: Prof. Dr.-Ing. Sören Hohmann
Korreferent: Prof. Dr. Tom Carlson

Impressum

KIT Scientific Publishing

Karlsruher Institut für Technologie (KIT)
KIT Scientific Publishing
Straße am Forum 2
D-76131 Karlsruhe

KIT Scientific Publishing is a registered trademark
of Karlsruhe Institute of Technology.
Reprint using the book cover is not allowed.

www.ksp.kit.edu

Print on Demand 2022 – Gedruckt auf FSC-zertifiziertem Papier

ISSN 2511-6312
ISBN 978-3-7315-1223-3
DOI 10.5445/KSP/1000148804

Preface

This thesis is the result of my work as a research assistant at the Institute of Control Systems (IRS) at the Karlsruhe Institute of Technology (KIT). Without the support of many people this work would not have been possible. First and foremost, I would like to thank Prof. Dr.-Ing. Sören Hohmann for providing me the research opportunity and for supervising my research. Furthermore, I express my gratitude for the inspiring discussions and the encouraging support in the past years. I would also like to thank Prof. Dr. Tom Carlson for his interest in my work and for the assessment of this thesis. I appreciated your participation in the evaluation committee and I enjoyed our conversations showing your genuine interest in my work.

Many thanks to the entire IRS staff for creating a pleasant working atmosphere. Especially, I would like to thank my colleagues Florian, Lukas, and Manuel, with whom I shared a room, for their support and for many constructive scientific and non-scientific discussions. Furthermore, I want to express my gratitude to the members of my research group, Balint, Christian, Florian, Jairo, Julian L., Julian S., Michael, and Sean for their support during my research, for proof-reading this thesis, and for providing valuable feedback. I am also grateful to all students who I supervised during their bachelor or master theses for their great support of my own research project. Furthermore, I acknowledge the fruitful collaboration with Dr. Manolis Chiou for experimentally evaluating my theoretical work.

Last but not least, I want to give many thanks to my friends for complementing my professional life and to my family for their unconditional and great support. Finally, Esther, I am grateful to have your support and honest feedback in my life, especially while writing this thesis, and for reminding me of the small, enjoyable things in life.

Karlsruhe, March 2022

Science is founded on uncertainty. Each time we learn something new and surprising, the astonishment comes with the realization that we were wrong before.

Lewis Thomas

Contents

List of Figures

List of Tables

Abbreviations and Symbols

Abbreviations

Abbreviation	Description
CMDI	Cooperative maneuver decision making interface
EMICS	Expert-guided mixed-initiative control switcher
GT	Automation design based on the game-theoretic n-stage war of attrition game model
GUI	Graphical user interface
HMC	Human-machine cooperation
LA	Automation design based on the leader-follower approach with the leader being the automation
LH	Automation design based on the leader-follower approach with the leader being the human
LOA	Level of automation
No.	Number
NASA-TLX	National aeronautics and space administration - task load index
NEMICS	Negotiation-enabled expert-guided mixed-initiative control switcher
NT	Automation design based on the negotiation-theoretic adaptive negotiation model
OCU	Operator control unit
POI	Point of interest
Q□	Question number □

Latin Letters

Symbol	Description
a	Action
A	Set of actions a
b	Evaluation function
c	Cost function
$C^{\square}$	Set of continously differentiable functions to the derivative order of $\square$
d	Decision option or direction
$d^{\square}$	Decision option or direction specified by $\square$ being a time instance / iteration number or a label
df	Degree(s) of freedom
D	Set of decision options d
e	Event
E	Set of events e
f	Probability density function
f_δ	Probability density function of utility differences δ
f_τ	Probability density function of thresholds τ
F	Cumulative distribution function of f
F_δ	Cumulative distribution function of f_δ
F_τ	Cumulative distribution function of f_τ
g	Error metric of EMICS
h	Hypothesis for Bayesian learning
$\boldsymbol{h}$	Vector of hypotheses for Bayesian learning
H	Set of hypotheses h for Bayesian learning
i	Placeholder for type of agent or player, either $\mathcal{A}$ or $\mathcal{H}$, i. e. automation or human
j	Placeholder for opposite type of agent or player compared to player/ agent i, either $\mathcal{A}$ or $\mathcal{H}$, i. e. automation or human
k	Running index, iteration number
l	Running index, stage index in the n-stage war of attrition
m	Stage index in the n-stage war of attrition
M	Mean value
n	Number of stages required in the n-stage war of attrition to reach an agreement
$n_{\square}$	Number of elements in sets or vectors $\square$ specified by index
N	Number of cooperation partners/agents/players
o	Offer
$o^{\square}$	Offer specified by $\square$ being a time instance /iteration number or a label
O	Set of offers o
$\tilde{O}$	Set of offers whose utility is greater or equal than current target utility
p	Probability
P	Set of players or agents

Symbol	Description
q	Probability offset for reinitializing hypotheses with probability zero
r	Risk disposition of agents
SD	Standard deviation
t	Time
u	Utility/utility function
$u^{\square}$	Utility specified by $\square$ being a time instance /iteration number or a label
u_{t}	Target utility function
U	Set of utilities/utility functions u
$\vec{U}$	Ordered utility set with elements in descending order
w	Weight
x	Auxiliary variable
y	Auxiliary variable

Greek Letters

Symbol	Description
α	Concession update step size or significance level
β	Adapation design parameter
γ	Effort function
δ	Utility difference
δ^{l}	Utility difference at stage l
δ^{m}	Utility difference at stage m
$\delta^{\dagger}$	Dirac function
Δ	Set of utility differences δ
ϵ	Concession rate
E	Set of concession rates ϵ
ζ	Importance level
Z	Set of importance levels ζ
θ	Parameter
$\boldsymbol{\theta}$	Parameter vector
Θ	Heavieside function
κ	Iteration number
λ	Type of player in incomplete information games
Λ	Set of types λ
μ	Specifier for decision options and offers
ν	Specifier for decision options and offers
π	Payoff/payoff function
Π	Set of payoffs/payoff functions π
σ	Variance of identification result
τ	Threshold/threshold function

Symbol	Description
τ^l	Threshold/threshold function at stage l in the n-stage war of attrition
τ^m	Threshold/threshold function at stage m in the n-stage war of attrition
T	Set of thresholds or threshold functions τ
ϕ	Transformation function
χ^2	Chi-squared distribution
χ^2_c	Critical value of chi-squared distribution
ψ	Game theoretic strategy
Ψ	Set of strategies ψ

Calligraphic and Other Symbols

Symbol	Description
$\mathbb{N}$	Set of natural numbers
$\mathbb{R}$	Set of real numbers
$\mathcal{A}$	Automation
$\mathcal{B}$	Bidding strategy
$\mathcal{C}$	Acceptance strategy
$\mathcal{E}$	Concession strategy or function
$\mathcal{F}$	Set of utility probability density functions f
$\mathcal{G}$	Game
$\mathcal{H}$	Human or $\mathcal{H}$-value of the Kruskal-Wallis test
$\mathcal{N}$	Negotiation
$\mathcal{T}$	Deadline
$\mathcal{J}$	Objective function
$\emptyset$	Empty set
∞	Infinity
$\mathcal{S}$	System

Indices, Exponents and Operators

Symbol	Description
$\square^*$	Optimal value
$\square^\diamond$	Distinct value
$\square^{\mathrm{H}}$	Set of history values
$\square^{-1}$	Inverted function
$\square^k$	Variable in iteration k or at discrete time t_k
$\square^\kappa$	Variable in iteration κ or at discrete time t_κ

Symbol	Description
$\square^l$	Variable at stage l in the n-stage war of attrition
$\square^m$	Variable at stage m in the n-stage war of attrition
$\square^+$	Set containing non-negative values
$\square^{>0}$	Set containing positive values
$\square^t$	Variable at time t
$\square_\mathrm{c}$	Critical value of a distribution
$\square_\mathrm{f}$	Final values of $\square$ at end of cooperative decision making
$\square_\mathrm{g}$	Variable with global scope
$\square_{\boldsymbol{h}}$	Function $\square$ with parameters of hypothesis $\boldsymbol{h}$
$\square_i$	Placeholder for kind of agent or player, either $\mathcal{A}$ or $\mathcal{H}$
$\square_j$	Placeholder for opposite kind of agent or player compared to i, either $\mathcal{A}$ or $\mathcal{H}$
$\square_\mathrm{l}$	Variable with local scope
$\square_l$	Running index
$\square_\mathrm{t}$	Target value or function
$\lvert\square\rvert$	Absolute value for scalars or number of elements for sets
$\arg\max$	Argument at maximimum
$\arg\min$	Argument at minimum
$\mathrm{d}\square$	Infinitesimal change of $\square$
$\Delta\square$	Describes difference between values of $\square$
$\frac{\partial\square}{\partial x}$	Partial derivative of $\square$ by x
$\dot{\square}$	Time derivative $\frac{\mathrm{d}}{\mathrm{d}t}\square$
$\frac{\mathrm{d}\square}{\mathrm{d}x}$	Total derivative of $\square$ by x
max	Maximization or maximum
min	Minimization or minimum
$\square^\top$	Transposed vector
$\square^{x:y}$	Cumulative value of $\square$ from stage x to stage $y-1$: $\sum_x^{y-1}\square$
$\bar{\square}$	Normalized variable
$\hat{\square}$	Estimated variable
$\tilde{\square}$	Constrained set

1 Introduction

Since the third industrial revolution in the second half of the 20th century, the automation of functionalities and processes, tools and technical systems has increased continuously and pervasively [BvD+14, PLF16]. To a great extent, the goal of this automation has been the state of full autonomy [Bai82, Lam13]. Nevertheless, up to the current day and despite all efforts to make human involvement redundant, industrial plants or vehicles require human operators or drivers, respectively, to supervise the automation's performance [End17]. Hence, these systems have become tools with widely automated functionality. The human has to interact with these tools only if necessary and to switch them on and off.

However, this form of interaction has some major disadvantages. One of them is the "out-of-the-loop performance problem" [End17] which describes the inability of humans to adequately react to a reduced performance of the automation. This is due to the fact that humans lack situation awareness in case they only possess a supervisory role [GDLB13]. Bainbridge summarized this and associated issues as the "ironies of automation" [Bai82]. A closely related disadvantage is the disregard of beneficial human action in situations in which the human outperforms the automation [VGLH11]. Another disadvantage is the costly development of fully automated systems which become increasingly complex and comprise more and more functions that have to be automated [Bil96, pp. 47-51].

To counteract these disadvantages, engineers started to focus on cooperative human-machine systems in the context of the fourth industrial revolution [EK99]. This implied reintroducing the human into the automated process or production and hence keeping her or him in the loop instead of allotting her or him a solely supervisory position [FWBB16, FDM+20]. Research has shown that human-machine cooperation creates performance synergies, e. g. by combining the strengths of the human (abstract thinking and situation recognition) and of the machine (endurance, consistent accuracy, and precision) [VGLH11], and increases human trust in and acceptance of technical systems [FCA+17, ACM+18, Fla19, NSWS20]. Furthermore, human-machine cooperation also allows for a step-by-step automation of a working technical system by gradually augmenting the degree of automation [PSW00]. Hence, engineers implement systems in which the human and the machine simultaneously share or sequentially trade control for a respective process or production task [PLF16, FCA+17, OGD17, ACM+18]. Examples are advanced driving assistance systems in partially autonomous vehicles [DvA+10, AMB12, LHFH18, Fla19, SWS19, WCW19], industrial production with close collaboration of human and machine [MLK+12, Lam13] and teleoperation of robots for search-and-rescue scenarios [CHS21] or surgery [RHS11].

All these examples and in fact the majority of current cooperative human-machine systems consider a close physical interaction of the human and the machine: The human and the machine operate on the same workpiece [MLM+11, MLK+12] or jointly control a vehicle [DvA+10, AMB12, Fla19, SWS19] or robot [NFA08, CHS21] by means of a steering wheel or a joystick. In all these cases, the human-machine communication is based on physical forces and haptic feedback (and potentially visual and acoustic feedback).

However, this form of communication allows only for a limited scope of human-machine cooperation as the interaction and interfaces are tailored to the specific use-case and application field, e. g. [ACM+18, Fla19, FDM+20]. The reason for this is the limited communication ability of haptic communication. One way to circumvent this problem is the development of supplementary communication channels such as brain-machine interfaces, e. g. [CD13]. However, these interfaces require significant technological effort, work to date only for specific brain signals in special cases, and are in general not yet user-friendly.

Moreover, a growing automation of tasks and processes entails an increase in the level of abstraction on which human and machine are able to communicate and interact [FAI+16, FWBB16, ACM+18]. This allows for richer communication symbols and ultimately for a larger scope of cooperative human-machine systems [ACM+18]. As a consequence, future cooperative human-machine systems with high degrees of autonomy require appropriate interaction design and foremost a holistic view on cooperation on higher levels of task execution [FAI+16, PLF16].

The next higher level of human-machine cooperation with respect to task execution is the so-called decision level [PLI15]: Current cooperative systems mostly *interact* [FWBB16, ACM+18] by e. g. cooperatively tracking given reference trajectories [NC15, LHFH18, Fla19]. Only a few approaches consider decision making scenarios during task execution. The vast majority of these approaches (implicitly) implements the leader-follower paradigm with the human as the sole decision maker, e. g. [GR86, SBP+18, TW19], or in form of decision support systems, e. g. [DvA+10, BAMF14, WWM+19]. Some approaches dynamically shift the decision making authority to the automation if its decisions are congruent to the human ones, e. g. [Khe11, MLK+12, MLH15, ABH+16]. However, in case of conflicting decisions the human remains the ultimate decision maker.

1.1 Towards Emancipated Cooperative Decision Making in Cooperative Human-Machine Systems

Implementing the leader-follower paradigm with the human in the lead has some disadvantages. Consider for example a highly automated driving scenario in which the vehicle's automation may possess more information about the future driving situation obtained by car-to-car communication. In this scenario, reasonable objections of the automation for maneuver selection may be ignored by the human if she or he is in the lead. Furthermore, the human may be left with too little information for decision making or too much information for processing. Both lead to an unfruitful interaction and potentially suboptimal decision making results. Similar concerns arise in the inverse scenario with the automation in the lead, e. g. if human perception of the immediate traffic situation outperforms the vehicle's situation recognition, e. g. due to blocked sensors.

To create synergies and to circumvent shortcomings of the leader-follower paradigm in the above example, it would be beneficial if the human had the ability to intuitively convince the vehicle's automation to follow her or his lead in e. g. maneuver selection. However, if the automation had good reason to disagree with the human choice of maneuver due to matters of e. g. safety, the automation should be able to communicate this in a comprehensible manner. This would lead to human and machine being engaged in an intuitive cooperative decision making process with equal rights and authority. Hence, human and machine would be *emancipated* cooperation partners. Furthermore, the process they were participating in had the objective to balance the significance of individual choices while treating both cooperation partners equally and to lead to a mutual agreement.

Therefore, if both cooperation partners are equally performant in terms of individual decision making and are able to participate in a cooperative decision making process, striving towards an emancipated human-machine cooperation on decision level offers benefits: in contrast to conventional leader-follower approaches, it allows to raise the synergies of cooperative decision making by means of information fusion or by cooperatively balancing and negotiating the significance of individual decision making. Furthermore, the equal assignment of authority within a cooperative setting has already proven to be beneficial by similar, successful concepts for human-machine cooperation on lower levels of task execution [NC15, Fla19]. Besides this, the equal assignment of authority within a cooperative setting does still allow for the generally applied paradigm that humans are able to switch off the automation.

To advance research on cooperative human-machine systems towards emancipated cooperative decision making, the objective of this thesis is the establishment of a first automation design enabled to participate in an explicit emancipated human-machine cooperative decision making and the evaluation of the automation design's potential benefits. For reasons of generalizability and reusability, the automation design should be model-based and should suit human concession behavior in cooperative decision making to increase user acceptance and trust.

1.2 Research Contribution

A first contribution of this thesis is a methodical classification of human-machine cooperation in Chapter 2 to precisely circumscribe the focus of this thesis. To this end, a new taxonomic model of human-machine cooperation, the *butterfly model* is introduced. Furthermore, Chapter 2 discusses existing literature on human-machine cooperation in terms of decision making in more detail and thereby reveals the corresponding research gap. The subsequently specified research questions in Section 2.4 are concerned with

1) suitably and mathematically modeling emancipated human-machine cooperative decision making processes focusing on human concession behavior,
2) adequately designing automation based on these models offering an intuitive interaction, and
3) appropriately evaluating and comparing new automation designs to state-of-the-art approaches by means of customized experimental designs targeting the cooperative decision making aspect.

To provide answers to those questions, the research of this thesis results in a first theory of emancipated human-machine cooperative decision making with emphasis on and consideration of human decision making and concession behavior. By means of the introduced mathematical models of cooperative decision making, automation designs are implemented and experimentally evaluated, demonstrating their practical relevance. In summary, the main contributions of the research reported in this thesis are therefore:

1) A first behavioral meta-model of emancipated human-machine cooperative decision making is introduced in Chapter 3, followed by the proposal of two mathematical behavior models originating from negotiation theory and game theory. The two novel mathematical behavior models aim to close the gap in terms of control authority between the two extremes of the leader-follower approach, i. e. the human in the lead has the ultimate control authority while the automation is only allowed to provide assistance and vice-versa. Both mathematical behavior models are theoretically analyzed and compared with respect to their ability to guarantee an agreement.
2) A study on the suitability of both mathematical behavior models to represent human concession behavior is reported in Chapter 4. Additionally and based on the proposed models, two automation designs are introduced and crucial aspects for their practical application are discussed.

3) A general experimental design focusing on human-machine cooperative decision making is established in Chapter 5 along with suitable measures to evaluate objective cooperative performance as well as subjective human perception. On this basis, two experimental evaluations of the proposed automation designs capable of human-machine cooperative decision making are presented in the same chapter. The experiments were conducted in the context of teleoperating a mobile robot with multiple levels of autonomy and guiding a highly automated vehicle. These experimental evaluations yield first evidence of the objective and subjective benefits of emancipated human-machine cooperation on decision level.

The resulting structure of the remaining thesis is depicted in Figure 1.1.

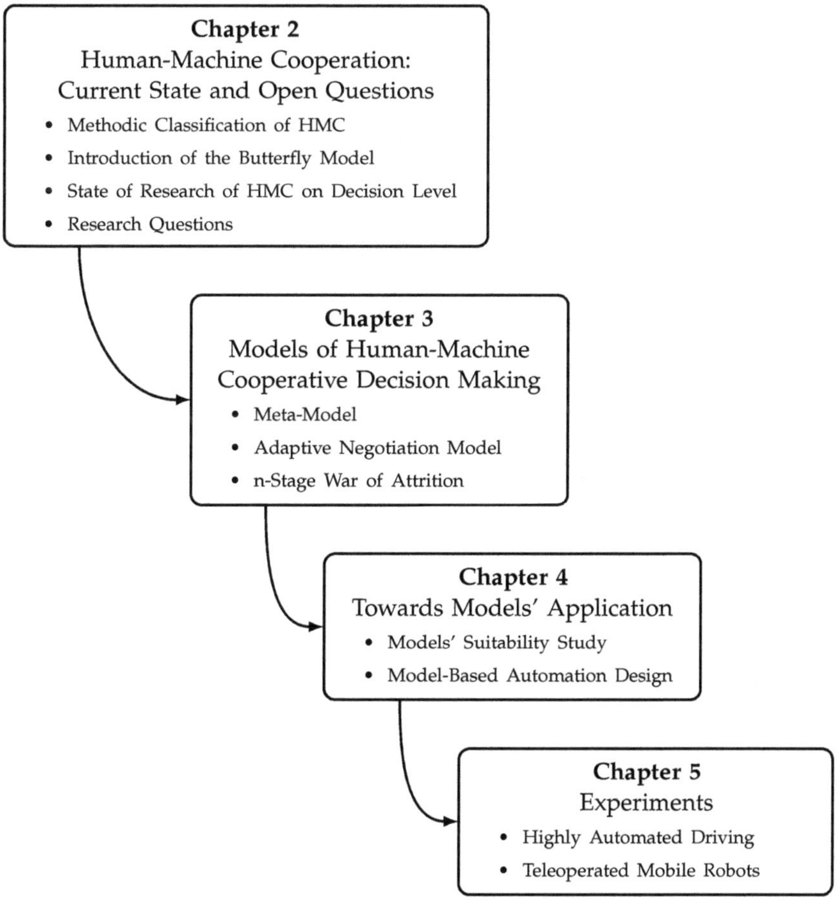

Figure 1.1: Structure of the thesis.

2 Human-Machine Cooperation: Current State and Open Questions

This chapter firstly introduces important terminology of human-machine cooperation in the context of this thesis in Section 2.1. For the purpose of circumscribing the scope of this thesis, i. e. human-machine cooperative decision making, Section 2.2 reports on the state of research of cooperative human-machine system design and provides a methodical classification of human-machine cooperation. For these purposes, the section presents an overview on good practice in terms of automation design for cooperative human-machine systems and elaborates on human behavioral models and their advancements towards models of human-machine cooperation by accounting for different interaction aspects. A review of existing human-machine cooperation models reveals some shortcomings with respect to classifying human-machine cooperation to the end of intuitively circumscribing the scope of this thesis. Therefore, a new taxonomic model, the *butterfly model*, is introduced. Upon this, Section 2.3 reports on research in the context of human-machine cooperative decision making and Section 2.4 reveals the open research questions that are addressed in this thesis.

2.1 Important Terminology

The following section discusses and defines important terminology for this thesis in the context of human-machine cooperation.

To start with, in this thesis *human* and *machine* denote the *agents*, i. e. active entities, in the considered interaction setting. For reasons of simplicity, this thesis considers only scenarios with one human and one machine. Whenever human and machine are put together they might interfere with each other and hence find themselves in a general setting called *human-machine interaction*. Note that *interfere* has no negative connotation in this context.

> **Definition 2.1 (Human-Machine Interaction)**
> *A general setting with two active entities, called agents, in which at least one of the agents (continuously) interferes with the other. One agent denotes the human, the other the machine.*

In the context of this thesis, the machine comprises an intelligence driving its actions, called *automation*. The design of this automation is within the scope of research reported in this thesis.

For a more detailed distinction of human-machine interactions, two obvious aspects are the distributions of *authority* and *ability* among the agents.

2.1.1 Authority vs. Ability

In this thesis, *authority* describes the *extent of permission/right* an agent possesses in the interaction or in parts of the interaction.[1] An agent with no authority has no right to interfere with others whereas the agent with the highest authority is in the lead. In other words, actions of agents are prioritized according to the agents' authority, e. g. actions of agents with little authority are only effective if agents with higher authority break down or have reached their goals. While the distribution of authority may generally be provided by nature or given by some sort of history, in the context of human-machine interaction, it is usually regulated by law giving a higher authority to the human, e. g. in case of driver assistance systems [WHLS16, Chap. 3]. However, in few cases, the machine is given a higher authority, e. g. in the application of electronic stability control systems in vehicles [WHLS16, Chap. 39]. Apart from this, the authority distribution can also be dynamically assigned, i. e. *shifted* or *traded*. Examples are authority shifts in driving assistance systems [FAC+03], in handover scenarios between human drivers and highly automated driving assistance systems [LHFH18], and in human-robot interaction [OKSB10, MLK+12, MLH15, KSB13]. Rarely, the human and the machine possess equal authority. Examples can be found in the development of fuel-saving driving assistance systems [Fla19] and in teleoperating mobile robots [CHS21]. Figure 2.1 presents an overview on the above discussed authority distributions in human-machine interactions. If interacting agents possess

Figure 2.1: Overview of authority distributions in human-machine interactions.

the same authority, i. e. they are equal in terms of authority, they are referred to as *emancipated*. Hence, this forms the basis of the following definition of *emancipated human-machine interaction*.

[1] Another closely related term to authority is *responsibility* which has, compared to authority, a notion concerning liability. However, this aspect is not in the research scope of this technically oriented thesis.

Definition 2.2 (Emancipated Human-Machine Interaction)
Consider a human-machine interaction according to Definition 2.1. If the human and the machine participating in this interaction possess equal authority, i. e. equal right to act, this interaction is called emancipated.

The emancipated human-machine interaction is in the focus of this work as motivated in the introduction of this thesis.

Note. *The proposition implied by Definition 2.2 targets human-machine interactions in which the functionality of the machine is secured and the emancipated interaction enables the creation of or increases synergies and mutual benefits. The general requirement that humans have to be able to switch off the machine is not affected by this proposition.*

In contrast to the authoritarian aspect, the *ability* focuses on the relation between an agent and a task and describes the extent to which an agent is able to solve it. Hence, agents with the *individual* ability to perform a task (or parts of it) can solve it (or the respective parts of it) without any help of other agents. However, this does not consider performance measures such as quality, efficiency, etc. In the context of human-machine interaction, it may be the case that both agents are not able to perform a task individually but can do so if they interact.

While there might be influences on each other, the aspects of authority and ability distribution in human-machine interaction can be regarded separately. Furthermore, the aspect of ability leads to two other sub-aspects that are crucial for this thesis and are discussed in the following: the ability for goal-oriented action (i. e. *rationality*, discussed in Section 2.1.2) and the general ability of the machine to perform a certain task (i. e. *level of automation*, discussed in Section 2.1.3).

2.1.2 Rationality

Rationality is a concept that describes to which extent an agent chooses its actions in a goal-oriented manner. Aggregating various definitions of rationality in literature on game theory [Moo85, FT91, SLB09] and discussions of human rationality [Nag95, CHC04, CGC06, CGIC09, YAB14, Str14, Har17, TLL^{+}18, AY21], this thesis applies the following definition of rationality.

Definition 2.3 (Rationality)
Agents act (fully) rationally when they strive towards a particular objective considering all potential influences of actions from themselves or others in the process of pursuing that objective. Agents exhibit a bounded rationality if they only consider influences of actions from themselves or others to a certain extent. Agents act irrational if the actively avoid reaching an objective.

In real-world scenarios, one usually has to assume bounded rationality for both, humans and machines, due to cognitive limitations (e. g. cognitive biases, limited thinking capacity, or time constraints [GV19]) or due to the complexity of the objective, see [Nag95, CHC04, CGC06, Har17, GV19].

In contrast to rationality applying to both human and machine, the level of automation explained in the following is an established measure for classifying the ability of machines.

2.1.3 Level of Automation

While machines outperform humans in some aspects such as strength and precision, humans are in general superior considering cognition and reasoning [VGLH11]. When interacting, it is crucial to be able to describe the extent to which the machine is able to perform on its own, without human support. This extent is generally referred to as the *level of automation (LOA)* for which literature offers various definitions, e. g. [End87, SLL78, PSW00, She11, BFH19]. Typically, these definitions are a set of level descriptions that divide the spectrum of performing a task by means of (human) manual control to full autonomy in discrete steps, see [SLL78, End87, EK99, PSW00, She11]. Additionally, Endsley and Kaber [EK99] and Parasuraman et al. [PSW00] enhance the LOA definition by introducing different LOA to different discrete "information processing stages", i. e. "acquisition, analysis, decision, and action" [PSW00], when performing a task. Apart from these discrete level definitions, Braun et al. [BFH19] define a continuous and quantitative metric to describe the LOA in human-machine interaction.

In this thesis, an exact (level) definition of LOA is not required. Therefore, the following broad definition based on the "criteria for LOA definitions" established by Braun et al. [BFH19] is applied.

Definition 2.4 (Level of Automation)
The level of automation (LOA) describes the extent to which the automation is (currently) acting autonomously. It ranges from manual control to full autonomy and is strictly monotone in between. The LOA may be associated with sequential and/or parallel aspects of human and machine jointly performing one or multiple tasks.

Remark. *Although LOA definitions usually originate from the ability of a machine to perform tasks or aspects of a task, the LOA is consequently linked to the authority of the machine to perform these tasks or task aspects, i. e. the machine will not be allowed to perform tasks or task aspects beyond its highest achievable corresponding LOA.*

With these aspects of human-machine interaction and corresponding definitions of rationality and LOA, human-machine *cooperation* can be defined.

2.1.4 Cooperation

General *cooperation* can be defined in various ways and domains (cf. biology [AH81], human-human cooperation also called *joint action* [SBK06], and automation design for human-machine interaction [BYK+02, FAI+16, FCA+17, BK17, Fla17, ACM+18]). One of the broadest definitions is given by Jean-Michel Hoc:

Definition 2.5 (Cooperation [Hoc01])
"Two agents are in a cooperative situation if they meet two minimal conditions.

1. *Each one strives towards goals and can interfere with the other on goals, resources, procedures, etc.*
2. *Each one tries to manage the interference to facilitate the individual activities and/ or the common task when it exists.*

The symmetric nature of this definition can be only partly satisfied." [Hoc01, p. 515]

Hence, cooperation requires an enhanced interaction in which agents strive towards an objective and interfere with each other to facilitate the achievement of this objective. Note that *facilitate* makes the difference between cooperation and *competition*.

In the following, the agents within a cooperation will be generally referred to as *cooperation partners*. Furthermore, depending on the modeling theory used to describe the cooperation partners, they are referred to as *agents* (also *automated agents* and *human agents*) or *players*.

A human-machine interaction fulfilling the requirements of Definition 2.5 is called *human-machine cooperation (HMC)*.[2]

> **Definition 2.6 (Human-Machine Cooperation)**
> *On the basis of Definition 2.5, human-machine cooperation is a human-machine interaction according to Definition 2.1 with agents, i. e. cooperation partners, possessing at least bounded rationality (see 1. in Definition 2.5 and Definition 2.3) and additionally each agent tries to to manage the interference to facilitate the individual activities and/or the common task when it exists (see 2. in Definition 2.5). The symmetric nature of this definition can be only partly satisfied.*

2.2 Methodical Classification of Human-Machine Cooperation

The following section aims for a methodical classification of human-machine cooperation to circumscribe the scope of this thesis. To this end, relevant literature on cooperative human-machine system design and on suitable classifiers is discussed. As a result, a new classifier in form of a taxonomic model for human-machine cooperation, called the *butterfly model*, is presented.

2.2.1 Introduction

The basis of today's research on human-machine cooperation was established in the second half of the 20th century by utilizing models of human behavior in the engineering context of so-called "cyber-physical systems" [Wie61]. Since then, a large body of literature has been created providing increasingly sophisticated human behavior models and their advancements towards models of human-machine cooperation. This also fueled the development of design paradigms for cooperative systems and corresponding automation designs for machines based on these developed models to interact and eventually cooperate with the human.

---

2 Another related term found in literature is *human-machine collaboration*. While some researchers define collaboration as an refinement of human-machine cooperation (e. g. collaboration enhances cooperation by the notion of actively working together or jointly performing tasks [BK17]), others do not seem to differentiate between these terms, cf. [Gro11, MLK$^+$12, FAI$^+$16, ACM$^+$18]. In this thesis, there is no need to differentiate between cooperation and collaboration. For reasons of uniformity, the term cooperation is used throughout this thesis.

In order to methodically categorize research on human-machine cooperation and to apply such a classification for circumscribing the scope of this thesis, the following three basic classifiers can be considered:

- **Number and Type of Cooperation Partners**
  The number and types (i. e. human or machine) of cooperation partners are the most basic classifiers for human-machine cooperation. However, in this thesis along with the vast majority of similar research, the focus is placed on the cooperation of one human and one automated machine, cf. Definition 2.1. Therefore, this classifier has low relevance and is not discussed further in this thesis.
- **General Aspects of Interaction**
  Depending on their abilities, authorities and the given interfaces, human and machine can interact within a cooperation in various forms, e. g. *sequentially* vs. in *parallel* or in *leader-follower*[3] form vs. in an *emancipated* manner.
- **Descriptive Behavioral Models**
  The behavioral models of cooperation partners in a human-machine cooperation originate from models of individual human behavior. These human behavior models comprise the human general abilities to act (in terms of cognition, reasoning, execution and learning) described from different perspectives such as psychology, ergonomics and engineering. Typically, these abilities are described on various dimensions and levels of abstraction.

The following sections elaborate on these classifiers by providing all necessary background information: Section 2.2.2 offers an overview of good practice in automation design for human-machine cooperation, followed by the explanation of existing human behavior models in Section 2.2.3, and of general interaction aspects in Section 2.2.4. Upon this background information, existing human-machine cooperation models which adopt (human) behavioral models for modeling both cooperation partners and enhance them by means of several interaction aspects are reviewed in Section 2.2.5. To counteract their shortcomings as classifiers for human-machine cooperation to emphasize the research focus of this thesis, a new taxonomic model, the *butterfly model*, is introduced in Section 2.2.6.

2.2.2 Overview of Good Practice in Automation Design for Human-Machine Cooperation

In the last decades, the increasing spread and pervasiveness of automation did not only yield a large variety of (partially automated) machines that do not continuously interfere with the human and are therefore tools for the human. It also enabled

---

[3] Another, equivalent term is *master-slave* which is avoided in this thesis due to the terminology's problematic historic background.

machines to perform certain tasks, such as driving or manufacturing, with at least temporarily or in parts comparable or even superior manners to the human. Due to the different strengths of humans (e. g. fast cognition and abstract thinking) and machines (e. g. physical strength, accuracy, computing power, and speed) engineers started to foster cooperative human-machine systems to benefit from the potential synergies and to cooperatively execute tasks better, safer, faster, etc. However, unlike conventional tool design aiming for the automation of basic functionalities for which suitable, mostly informative human-machine interfaces are required to achieve high usability, cooperative human-machine systems pose a greater challenge. This is due to the fact that a lot of automation design effort is required to suitably manage the interaction with the human to achieve the targeted benefits of the cooperation. The interaction management has to consider many aspects which include taking into account human behavior, i. e. learning and adapting to it, completing the given task in cooperation, assuring safety of the human, assisting and supporting the human and handling conflicting interests. In other words, it requires much effort to turn the static automation design of tools into dynamic, adaptable automation designs.

Due to ethical and legal reasons (a comprehensive overview is provided by Flemisch et al. [FDM+20]), most of the research on automation design aiming for a successful cooperation of human and machine follows *human-centered* design approaches. They focus mainly on the human needs, abilities and behavior and on how machine interaction may have a positive impact.

Two prominent design concepts for the automation in human-machine cooperation are the concepts of *traded and shared control*, in which the cooperation partner *sequentially trade* or *continuously share* the authority of conducting a task in cooperation. These concepts usually define cooperation partners to be (at least temporarily) equally capable of individually performing the task in question [ACM+18]. Especially in the case of the term "shared control", there are many slightly different definitions in literature revealing the lack of unity among the peer researchers, cf. [EK99, PLI15, FAI+16, Fla17, ACM+18]. One major reason for this issue is the large range of scopes and applications of cooperative human-machine systems, e. g. in medical technologies, driving assistance systems, and robotics [ACM+18].

While the above and similar concepts offer guidelines for human-centered automation design considering the abilities of human and machine and their authority in interaction, other concepts focus on the human behavior and reasoning. One prominent example is the concept of *mental models* which were first extensively discussed in the eponymous book by Gentner and Stevens [GS83]. Humans form mental models of everything they encounter: the world, other people, and technical systems. By means of these models, humans are able to "predict system behavior and guide actions" [Nor83]. Together with their peer researchers, Norman, Gentner and Stevens [GS83, Nor83] early highlighted the necessity to properly take into account human mental models in system design to develop appropriate human-machine interfaces. Subsequently, Heiner Bubb [Bub03] postulated that the human is naturally developing and utilizing mental models of the machine she or he is interacting with. Furthermore, he assumed that the mental models have to correspond with reality to a certain extent such that human-machine interaction is beneficial and "human errors" can be avoided. In order to achieve this correspondence despite the missing ability to identify mental models of humans, he introduced the *system ergonomics approach* enabling designers to find and implement the "simplest form of operation" [Bub03]. Flemisch et al. [FSKL08] promoted mental models in the context of human-machine cooperation and proposed design guidelines to ensure the *compatibility* of human mental models of the machine the human is interacting with and the behavior models of the machine. However, this compatibility requirement does not imply similarity in behavior of human and machine. It rather demands for automation designs such that the human is able to establish a mental model of the automation. As a result, humans are able to predict the automation behavior and will not face an uncomfortable nor uncertain situation. Nevertheless, adopting human behavioral models for the automation design in cooperative human-machine systems is assumed to increase compatibility of human mental models and corresponding automation behavior and to ultimately lead to a more successful cooperation between human and machine [FSKL08]. In other words, designing the automation in accordance to human models is supposed to result in interactions between human and machine that are less disruptive, increase human acceptance and yield greater cooperative performance.

Following this concept of replicating human behavior by designing automation accordingly, researchers have two potential approaches to develop a model of human-machine cooperation. These approaches are depicted in Figure 2.2: Starting from human behavioral models, the first approach adopts the insights on human behavior in an automation design for human-machine cooperation which supports and seamlessly adapts to the human (dotted arrow in Figure 2.2). Alternatively, the second approach advances the human behavioral models to human-human cooperation and then transfers these models to human-machine cooperation (dashed arrows in Figure 2.2). Although the latter approach tackles the fact that human behavior changes in cooperation [IFH19], most researchers follow the first approach to establish models of human-machine cooperation [FSKL08, FBB+14, PLI15, ACM+18]. However, these models resulting from the first, direct approach usually assign implicitly a higher authority to the human compared to the automation, see e. g. [ACM+18]. In contrast to this, human and machine possess equal rights from a modeling perspective in case human-machine cooperation models are established following the second approach considering emancipated cooperation partners.

In summary, the good practice of automation design for human-machine cooperation is a collection of guidelines and principles that highlight the importance to consider human needs, abilities, behavior, reasoning and mental models. Researchers accounted for this by establishing models of human behavior and advancing them to behavior models of partners in a human-machine cooperation. In what follows,

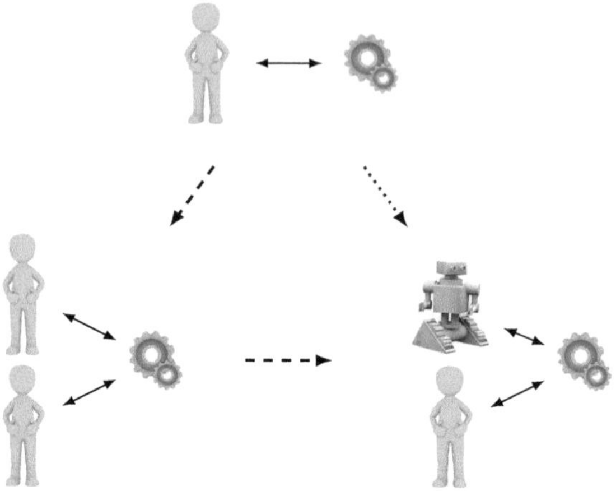

Figure 2.2: Abstract representation of the two different approaches to develop human-machine cooperation models starting from individual human behavioral models: the direct approach indicated by means of the dotted arrow and the approach via human-human cooperation, hence considering emancipated cooperation partners, shown with dashed arrows.

this development is elaborated on, starting with the report on models of human behavior.

2.2.3 Cognition, Reasoning, Execution and Learning in Human Action

In the early years of the second half of the 20th century, psychologists agreed that human social behavior is *goal-directed* (e. g. [Hei58])[4], i. e. human action follows some sort of plan [Ajz85]. To explain the origin of this plan, the psychologists Fishbein and Ajzen introduced the *theory of reasoned action* [FA75, AF80] for predicting human social behavior in situations in which humans are able to willingly control their actions. According to this theory, humans consider available information and predict the implications of their actions. This process forms an intention to perform an action which in turn leads to the action itself if no unforeseen events occur. Fishbein and Ajzen later refined this theory with respect to the determinants of the intentions in order to cover also situations in which humans (anticipate to) possess no full control over potential actions. This resulted in the *theory of planned behavior* [Ajz85]. Both theories are based on experimental data and were also experimentally compared which proved that the theory of planned behavior enhances the theory of reasoned action [MEA92].

4 This insight is also backed up by the research on sensorimotor control of human actions that has been proven to be optimal with respect to some goal [Fri11].

Upon these general insights from the field of psychology, engineers[5] started to develop detailed models of human behavior, see [Don82, Ras83, Mic86]. Although these models are usually based on some experimental evidence for some of their features, the overall models are typically not validated, e. g. [Don82]. The focus of these models was to appropriately design automation to suit human behavior in various aspects such as interface or assistance design. Most literature addressing human behavior from the engineering perspective considers the *cognition-and-action cycle* (also known as the *perception-action cycle*) with the following typical elements: *cognition* of the general current situation, human *reasoning*, i. e. processing the obtained information and deriving potential future action, and *execution* of the determined action. With respect to different aspects described in human behavior models, the cognition-and-action cycle is typically defined with different levels of abstraction.

One of the first human behavior models is the work of Jens Rasmussen [Ras83]. He introduced three levels to describe the behavior of a skilled operator in a deterministic environment. In essence, each level in this model defines human behavior as a cognition-and-action cycle with respect to a certain *degree of consciousness*. Depending on the task complexity, its frequency of occurrence and degree of consciousness during execution, human behavior is goal-driven and either *knowledge-*, *rule-* or *skill-based*:

- **Skill-Based Behavior**
  This level describes *sensorimotor performance* of humans during activities following some intention without conscious control. Such behavior is associated with often performed and well trained tasks. On this level, the *sensory input* is converted into *signals* that directly trigger *automated sensorimotor patterns*. Therefore, behavior on this level can be compared to feedforward control or feedback control if error information is available.

- **Rule-Based Behavior**
  The behavior on this level is for tasks for which some experience is available. However, the tasks still require conscious attention: The human *recognizes* which *task* is appropriate based on *signs*. This task is associated with *rules* which are established by experience and appropriately compose the execution of automated behavior patterns of the skill level.

- **Knowledge-Based Behavior**
  In unknown situations, human behavior to reach a known goal consists of the *identification* of the situation on a *symbolic* basis and the *decision* on the right task to reach the known goal which involves *planning* and validating by trial and error or by predictions.

Note that this model explicitly considers *learning* and training effects which will shift task execution towards skill-based behavior. Furthermore, humans can also

[5] In the following, this thesis focuses on the engineering perspective of human-machine cooperation models, i. e. their practical application in the automation design for human-machine systems.

actively focus on task execution e. g. due to unknown circumstances which will shift it towards a knowledge-based behavior.

In contrast to Rasmussen's work focusing on the degree of consciousness with which humans perform, Edmund Donges [Don82, Don99] chose another approach that is centered around the *degree of task abstraction* and is specified for the task of driving. The resulting model possesses three levels: On *navigation level*, a suitable route from the starting position to the known destination with respect to a corresponding time schedule is determined. The *guidance level* refines the route and time schedule and provides reference trajectories that include the desired car velocity and respect current local traffic. Up to this point, the model postulates an open-loop control for the cognition-and-action cycle. This changes in the lowest level, the *stabilization level*, on which the reference trajectory is supposed to be tracked by means of closed-loop control concepts. Although Donges and Rasmussen chose different focuses for their models, Donges proposes a mapping of the two level models in [Don99]: Navigation is associated with knowledge-based behavior and stabilization corresponds to skill-based behavior. Guidance may be associated with either of the three levels of Rasmussen depending on the experience of the driver.

Another similar example of modeling driver behavior was proposed by Michon [Mic86] who divided the driving task into three levels: On the *strategical level* (also planning level), the destination and the general route with corresponding risks and costs are derived. On the *tactical level* (maneuvering level) drivers determine appropriate driving maneuvers such as turning and overtaking which have to be in accordance with the derived plan from the strategical level. On the *operational level* (control level) the chosen maneuvers are instantiated. Depending on the maneuver execution, there is the possibility to adapt the maneuver choice and also the strategical plan if required.

In more recent work in the context of LOA research, human behavior models elaborated on the perception-action cycle of human performance to define aspects which can be supported or conducted by the automation. Considering the increasing automation of human-machine systems, Endsley [End17] aggregated work of Endsley and Kaber [EK99] and Parasuraman et al. [PSW00] and described three stages of task performance: possessing *situation awareness*, making a *decision* on potential actions and performing this *action* [End17]. Similar to this and to the work of Parasuraman [PSW00] is the work of Pacaux-Lemoine and Itoh [PLI15] which proposes similar stages: *information gathering*, *information analysis*, *decision making* and *action implementation*.

In summary, the existing human behavior models in literature were proposed along three prominent dimensions. The first dimension is concerned with the *perception-action cycle* of humans and is sometimes referred to as the *horizontal* dimension. The second dimension deals with the degree of *task abstraction*, sometimes called the *ver-*

tical dimension, and is greatly inspired from an engineer's design perspective.[6] The third dimension is associated with the degree of *consciousness* which is greatly influenced by learning effects. The relation between these dimensions is depicted in Figure 2.3. Note that the second and third dimensions are not motivated by strong experimental evidence. They are purely motivated by observations as they serve an engineering purpose. Furthermore, note that existing mathematical models typically do not consider all dimensions nor all potential levels associated with each dimension, e. g. optimal control models of sensorimotor control focus on the entire perception-action cycle but only on the lowest level of task abstraction and neglect the dimension of consciousness/learning, see [Fri11]. For reasons of better readability, the term *(behavioral) level* refers to task abstraction level in the following if not specified otherwise.

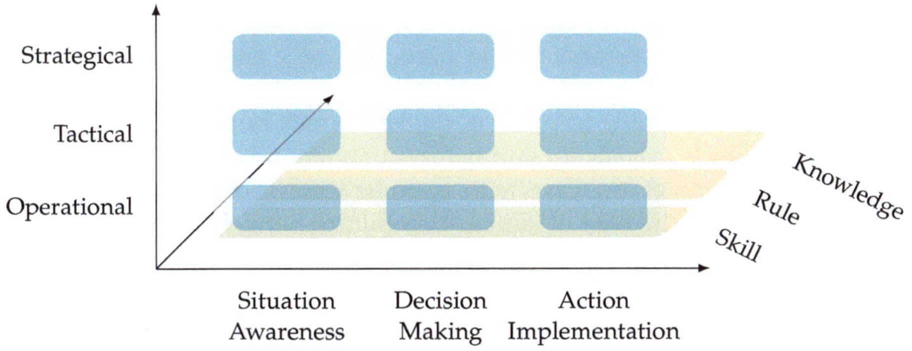

Figure 2.3: Dimensions of human behavior models: perception-action cyle (horizontal), task abstraction (vertical), consciousness (depth); aggregated from [Ras83, Mic86, End17]. The colored boxes abstractly illustrate levels and components of actual human behavior models.

Around the same time the human behavioral models discussed above were introduced, psychologists came up with the concept of mental models that humans establish of everything, and especially of other humans and technical systems they encounter, to understand and predict potential interaction with them and resulting consequences [GS83]. Following research has shown that humans need to be able to establish such mental models of technical systems in order to successfully interact with the technical systems, see [Nor83, FSKL08] and Section 2.2.2. Consequently, engineers developing cooperative human-machine systems should apply human behavioral models within the automation design. To this end, they established models of human-machine cooperation which adopt (human) behavioral models of the cooperation partners. Furthermore, these models account for other general aspects of the interaction which are discussed in the following.

6 The influence of conventional automation design on human behavior models becomes apparent regarding the hierarchical design concept for the automation of complex systems, see e. g. [Sar83, Bro86, VNE+01].

2.2.4 General Aspects of Interaction

The general aspects of interaction[7] within human-machine cooperation are the *timing* of the interaction and the *ability* and the *authority* of the cooperation partners.

Regarding the aspect of timing, the cooperation partners can either interact *sequentially*, i. e. alternatingly, or in *parallel* depending on the given interface and task. In sequential interactions, one cooperation partner acts first followed by the other one, e. g. the automation proposes actions for completing a task and the human chooses one to be implemented [MM95]. Parallel interaction can be often found in haptic human-machine cooperation, e. g. in the case of human and assistance system simultaneously controlling and hence influencing a vehicle [NC15, FCA$^+$17, ACM$^+$18, Fla19, IFH19].

The aspect of ability considers cases in which human and machine possess *complementary* capabilities to perform certain parts of a task and therefore require cooperation to complete the overall task. Furthermore, situations in which human and machine are both capable to perform the entire task but cooperate to *share* the workload or to increase *redundancy* and hence safety are taken into account as well. Schmidt [SRBL91] denotes the case with complementary capabilities as "integrative" and distinguishes the case of similar capabilities between "augmentive form" (workload is shared by allocating sub-tasks to the different cooperation partners) and "debative form" (the workload is not shared, each cooperation partner performs the task individually and the outcomes are debated). In the same context, Pacaux-Lemoine [PLD02] proposed to enhance the term of human abilities to not only comprise abilities to individually operate but also the abilities to cooperate:[8] Denoting the dimension of human abilities to operate (including the perception-action cycle with the elements of *information gathering, information analysis, decision making,* and *action implementation,* see also Section 2.2.3) as the human *know-how* (to perceive and act), they named the human abilities to cooperate the *know-how-to-cooperate* consisting of the operational elements *information gathering on the other, detection of interference, management of interference* and *function allocation* [PLI15]. The latter element determines which form and degree of cooperative task execution (e. g. shared vs. integrative) is applied in a given situation.

In close relation to the ability of the cooperation partners, the aspect of authority within cooperation possesses a key role in cooperative system design. Obviously, a cooperation partner with a limited capability to perform tasks or parts of a task is also accompanied by a limited authority in performing cooperatively. Traditionally, such limitations are associated with the machine. Additionally, other reasons based on law and (re-)liability often lead to a reduced authority of the machine

[7] These general aspects are at first independent of any potential behavioral level of the cooperation partners. Furthermore, if different behavioral levels are considered, the manifestation of the interaction aspects may differ across these levels.

[8] On this basis, Pacaux-Lemoine also defines *levels of cooperation* similar to LOA [PLV13].

within the cooperation [FDM+20]. As a consequence, there are typically two forms of authority distribution among the cooperation partners: the automation has no authority to execute actions and is left in an *assistive* role, supporting the human who has all execution authority (leader-follower paradigm). In the other case, human and machine *share* and/or dynamically assign the authority within the cooperation, e. g. [MLK+12, Fla19]. Millot and Mandiau [MM95] denote these cases of assistance and authority sharing by "vertical" and "horizontal" cooperation. In an untypical third form of authority distribution, the automation has all authority, e. g. due to its learning abilities with respect to human behavior (denoted as "implicit mode of cooperation" by Greenstein et al. [GAR86]).

Figure 2.4 summaries and depicts the different categories of human-machine cooperation along the aspects of timing, ability and authority describing the general form of interaction.

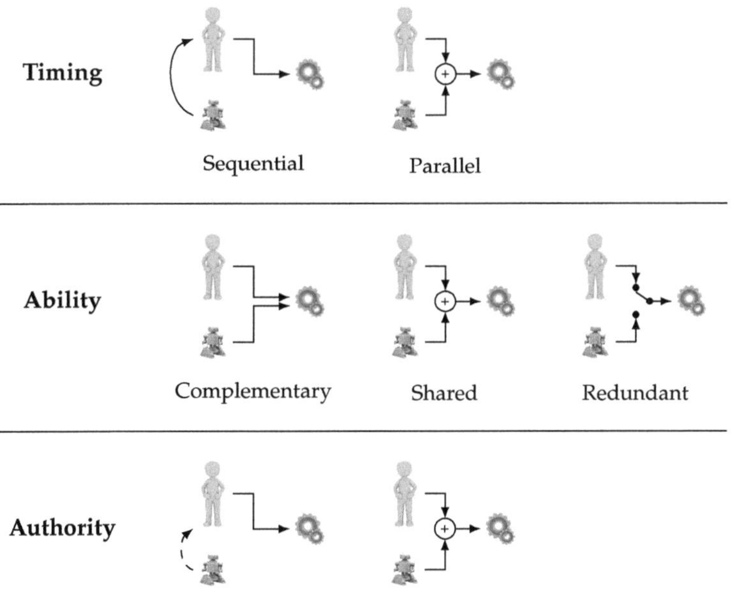

Figure 2.4: Forms of interaction within human-machine cooperation considering the aspects of timing, ability and authority. Arrows indicate the course of action. In case of sequences, i. e. for sequential and assistive forms of interaction, only one variant is depicted. Perception aspects are neglected in this overview. Partially inspired by [PLF16].

Upon the introduced models of human behavior and the general aspects of interaction, the next section discusses layer models of human-machine cooperation for the purpose of classifying research in the context of human-machine cooperation and circumscribing the scope of this thesis.

2.2.5 Layer Models of Human-Machine Cooperation

In order to properly design cooperative human-machine systems and especially the automation within, the modeling of the overall human-machine cooperation has proven to be methodically beneficial: Flemisch et al. [FSKL08] expect a more successful cooperation if there is a compatibility of the mental model of the automation behavior developed by the human and the automation behavior itself. To achieve this, behavioral models of the human can be advanced towards models of human-machine cooperation. This is often accomplished by introducing behavioral models for the automation design which resemble the model of human behavior, see Section 2.2.2. For models concerned with general human-machine cooperation, this implies a mirroring of the human behavior models typically based on task abstraction levels (see Section 2.2.3) for the automation behavior. The result are layer models of human-machine cooperation. The following paragraphs provide an overview on the existing layer models of human-machine cooperation.

Flemisch et al. [FSKL08] proposed a layer model of human-machine cooperation in the context of cooperative vehicle control. Within this model they aggregate the vertical (task abstraction) and horizontal (perception-action cycle) dimension of human behavior models [Ras83, Don99, EK99, PSW00] and adopt the so-developed human model in large parts for the automation behavior modeling. This results in two almost identical behavior models of human and machine which cooperatively interact with the vehicle. Both behavior models comprise a perception module and a situation assessment module to perceive and assess the state of the vehicle and the environment it is in. This is followed by a four layer reasoning model describing the task of controlling the vehicle with four levels of abstraction. The four levels are closely related to the human behavior model of Donges [Don82, Don99]: On the *navigation level* a route is planned to reach the destination. The *maneuver level* decides on meaningful maneuvers that suit the predefined route. Each maneuver is converted into a trajectory on the *short term planning level* and finally into control actions on the *control level*. The control actions of human and automation are then combined via human *interaction resources* and an *arbiter* module of the automation. The arbiter's objective is to resolve conflicting actions of human and automation via some arbitration process. Furthermore, the interaction model allows for different degrees of automation such that the participation of human and machine in action execution does not have to be equal. The authors also point out that the cooperative control loop shall be closed on all four levels simultaneously. Together with the replicated human behavior in automation design, the authors assume that the automation presents a human *compatible* behavior and hence leads to better interaction and cooperation.

In their subsequent layer models, Flemisch et al. [FBB+14] focused on the actual human-machine interaction in terms of communication on each level of the vertical dimension of the driving task abstraction. To this end, they reduced the number of task abstraction levels to three (*navigation, guidance,* and *control,* similar to Donges [Don99]) while the guidance level is split into *maneuver* and *trajectory* guidance. On this basis, they discuss parallel and serial aspects of cooperative vehicle guidance and control: Human and automation may navigate, guide and control in parallel according to the current degree of automation which depends on the capabilities of human and automation. The automation *displays* results of the different levels and the human is able to *intercept* on all levels (cf. "mediator" concept in [BAMF14]). Consequently, human and machine communicate on all levels but the human has the ultimate authority and the automation possesses an assistive role. Providing the concept of *steer-by-wire,* the researchers also highlight a sequential aspect of the cooperation which is closely related to the LOA: the automation may take responsibility of guidance and control whereas the human mostly focuses on the navigation. The LOA may be adapted dynamically depending on automation capability and human focus. Shortly after this publication, Flemisch et al. [FAI+16] generalized the scope of their model and introduced new names for the levels of task abstraction: *strategic, tactical,* and *operational.*

Pacaux-Lemoine and Itoh [PLI15] proposed a layer model of human-machine cooperation considering the vertical and horizontal dimension of human behavior models for a generic scope: the three vertical levels of task abstraction are denoted as *planning, tactical,* and *operational.* Furthermore, Pacaux-Lemoine and Itoh focus on an enhancement of the horizontal perception-action cycle of a human towards human capabilities of cooperating, i. e. *know-how* (to perceive and act) towards *know-how-to-cooperate,* see Section 2.2.4. Consequently, these human capabilities are then also introduced to the automation model. Additionally, the capabilities to cooperate influence the "mixing (or not) of [...] results" [PLI15] of the conventional horizontal perception-action cycles of human and automation: human and automation may e. g. analyze information cooperatively or one of them does and shares the results. The concrete assignment and result sharing depends on the cooperation partners' interaction/communication capabilities, analyzing capabilities and workload [PLI15]. The close relation of the cooperation models of Flemisch et al. [FSKL08, FBB+14] and Pacaux-Lemoine and Itoh [PLI15] are discussed in a joint publication of the corresponding researchers [PLF16].

Abbink et al. [ACM+18] introduced a layer model of human-machine cooperation with a generic robotic scope comprising four "task levels" (*strategic, tactical, operational,* and *execution*) for each cooperation partner. Between these task levels, the model assumes a "goal sharing/multi-modal communication interface" to transform the result of a higher level (called "action") into a "goal" for the next lower level. These goals can also be shared/traded with the cooperation partner. The authors do not elaborate on the nature of these interfaces. Each task level has access to a "multi-sensory channel" to perceive the environment and the system and to assess the task progress. Furthermore, the model includes at each task level the degree of consciousness (i. e. skill-, rule-, and knowledge-based behavior, see [Ras83] and Section 2.2.3) of each cooperation partner to account for the aspect of learning behaviors. Consequently, communication between cooperation partners via a "multi-modal interaction interface" for each task level has to suit the partners' current degrees of consciousness on the specific task level. The authors point out the advantage of this integration in terms of modeling simultaneous guiding (i. e. "teaching") and learning which is assumed to be beneficial for a "symbiotic relationship" between human and machine.

Flemisch et al. [FAI+19] enhanced their previous layer model [FAI+16] which possessed a generic scope and the three task abstraction levels *strategic, tactical,* and *operational* by means of highlighting the aspects of cooperation on higher levels. To this end, the model comprises a meta layer for communication among the cooperation partners, called "cooperational" [FAI+19], transversal to the task abstraction levels. By means of this layer, the authors accounted for the postulated *know-how-to-cooperate* of Pacaux-Lemoine et al. [PLD02, PLI15]. Therefore, this layer may include "communication about the cooperation" [FAI+19] and resembles the model's new focus on the communication on all levels of human-machine cooperation. Furthermore, the authors discussed the close relation to and integration of the above introduced model of Abbink et al. [ACM+18].

Table 2.1 provides an overview of the discussed layer models of human-machine cooperation along the following features: the levels of task abstraction, the stages of the considered perception-action cycle, and the consideration of the cooperation aspect.

In summary, existing layer models of human-machine cooperation have evolved from duplicating and slightly adapting human behavior models based on task abstraction levels to models that increasingly consider the aspect of cooperation on all these task abstraction levels. Furthermore, existing layer models differ in some aspects due to different scopes, origins, modeling focuses, and despite the clearly noticeable will of researchers to align their models.[9]

Apart from being well-motivated, all of these models lack evidence for the existence of the postulated layers. Furthermore, when taking a closer look at the concepts and approaches associated with the discussed layer models of human-machine cooperation, they are either:

- General design concepts for human-machine cooperative systems (e. g. "H-metaphor" in [FAC+03], "H-mode" in [FBB+14, ABC+16], "AiKiDo metaphor" in [FPLV+20], all associated with the layer model of Flemisch et al. [FBB+14, FAI+16, FAI+19]),

[9] The struggle to align models is most noticeable in the researchers' discussion of the relation of the design paradigm *shared control* and state-of-the-art layer models of human-machine cooperation: While Flemisch et al. [FAI+16, FAI+19] described shared control as being mostly applied on the operational/control level of human-machine cooperation, some of the authors advanced the term *shared control* to also comprise all layers of human-machine cooperation [ACM+18].

Table 2.1: Overview of most relevant layer models of human-machine cooperation.

Layer Model	Levels of Task Abstraction	Perception-Action Cycle	Aspects of Cooperation
[FSKL08]	navigation maneuver short-term planning control	present	arbiter concept
[FBB$^+$14]	navigation guidance control	not explicitly present	mediator concept
[PLI15]	planning tactical operational	present	know-how-to-cooperate
[ACM$^+$18]	strategic tactical operational execution	not explicitly present	multi-modal interaction interface & goal-sharing/multi-modal communication interface
[FAI$^+$19]	strategic tactical operational	not explicitly present	cooperational layer

- Descriptive concepts of information exchange within human-machine cooperation (e. g. "know-how-to-cooperate" in [PLD02] associated with the layer model of Pacaux-Lemoine et al. [PLI15], "interaction patterns" in [BLF19] associated with the layer model of Flemisch et al. [FAI$^+$19]), or
- Implemented approaches for the automation design in human-machine cooperation (e. g. decision support in driving assistance systems [DvA$^+$10] associated with the layer model of Abbink et al. [ACM$^+$18], "mediator" concept in [BAMF14], "self-determined nudging" in [WAS$^+$19], both associated with the layer model of Flemisch et al. [FAI$^+$19], and "emulated haptic feedback brain-computer interface" in [PLHSC20] associated with the layer model of Pacaux-Lemoine et al. [PLI15]) which do not consider all levels and dimensions of the layer models and typically not the emancipated interaction of the human and the machine.

Consequently, there are no implemented approaches which comprise the entire scope of any layer model of human-machine cooperation. Hence, the existing layer models serve two major purposes:

- The layer models are a means for researchers to communicate and highlight certain aspects of human-machine cooperation which are in their interests or focuses.
- The layer models are of structural value to classify the researchers' work on human-machine cooperation.

With regard to the topic of this thesis, i. e. emancipated human-machine cooperative decision making, none of the above discussed layer models allows for an intuitive communication and a clear classification of the thesis' research: decision making is typically associated with each level of task abstraction and with the perception-action cycle. Given these observations and pursuing the objective to circumscribe the research reported in this thesis, a new taxonomic model, the *butterfly model*, was introduced: it was established from an engineering perspective to structure and relate existing work on emancipated human-machine cooperation and to circumscribe the research on emancipated human-machine cooperative decision making reported on in this thesis.

2.2.6 Butterfly Model of Human-Machine Cooperation

The taxonomic model of human-machine cooperation introduced in this section is called the *butterfly model*. It was established in the course of two supervised master theses [Sch18, Ste18] and published thereafter [RWIH20]. The butterfly model is defined from an engineering perspective on how to *executing a general task* with focus on the aspects of *emancipated cooperation on all levels* of task abstraction. The result is a *lean taxonomic model* which is inspired by the layer models of human-machine cooperation (see Section 2.2.5) and which allows to structure and relate existing implemented work and the approach of this thesis on emancipated human-machine cooperation.

Introduction of the Butterfly Model

The butterfly model[10] is depicted in Figure 2.5 and will be discussed in detail in the following.

The key features of the butterfly model are:

- No constraints, also no implicit ones, on the authority distribution among cooperation partners which allows for a potentially emancipated cooperation between human and automation.
- No constraints on the application scope, i. e. the model is verbalized for the generic case of *task execution*.

[10] The name of the butterfly model is inspired by its shape.

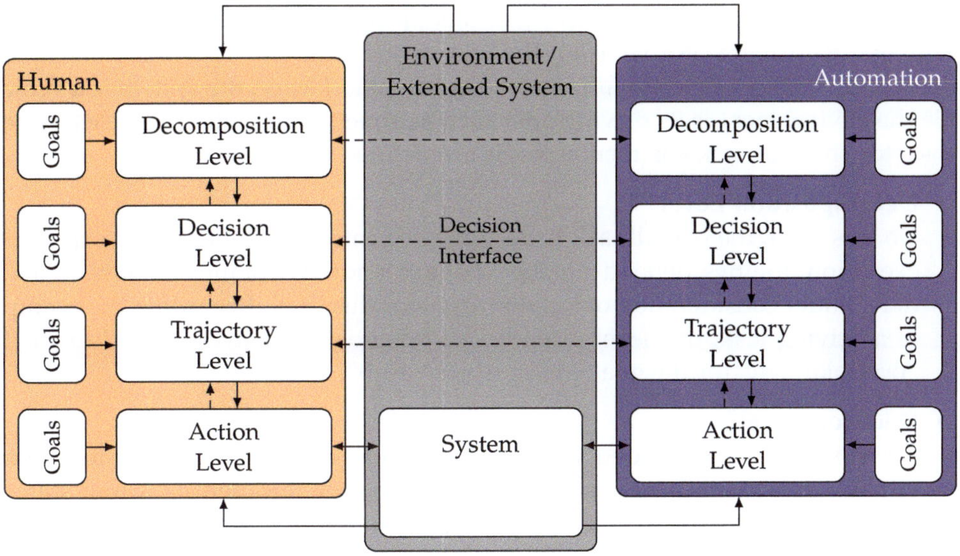

Figure 2.5: The butterfly model of human-machine cooperation inspired by the model of Flemisch et al. [FBB+14, FAI+16] but with focus on interfaces for an emancipated goal-directed cooperation between human and automation on every level.

- Focus on interconnected task abstraction levels within the model of one cooperation partner for reasons of representation simplicity, while integrating the elements of *decision making* and *action implementation* of perception-action cycles in the task levels and disregarding different degrees of consciousness, i. e. learning aspects.
- Potentially individual goals for each cooperation partner and specific objectives for each level.
- Explicit possibility to directly communicate, interact and cooperate between cooperation partners on all task abstraction levels via suitable interfaces. This also allows for an easy representation of systems with increased or dynamically changing LOA.
- Taxonomic model of human-machine cooperation with layers which are placeholders for more specific models of human-machine cooperation.

The *human*, the *automation* and the *environment* form the fundamental elements of the butterfly model. Both, human and automation, are able to perceive the environment. Within the environment, there is a *system* the human and the automation primarily interact with, e. g. a vehicle or a work piece. Its state is observable for both human and automation.

In the following, the task abstraction levels of both cooperation partners are defined in more detail. Although, the scope of this model is not limited and may cover various applications (e. g. in cooperative manufacturing involving humans and robots, or cooperative driving of a vehicle), the task abstraction levels are exemplarily explained with respect to the execution of a driving task. Hence, the system is a vehicle while the environment is its driving area such as streets, cities, other vehicles, pedestrians, etc. The four task abstraction levels are defined as follows:

- **Decomposition Level**
  On this level, the overall task is decomposed into all potential subtasks whose execution abilities depend on the system's and environment's state. This is done under consideration of a certain goal for this level. Regarding the example of a driving task, this level provides all potential maneuvers, e. g. "turn left", "overtake", etc. for the goal "drive from A to B".

- **Decision Level**
  On this level, it is decided which subtask, i. e. driving maneuver, to execute with respect to the system's, i. e. vehicle's, and environment's current state as well as given objectives like task execution in shortest time or with the least effort, i. e. with minimal travel time or steering effort. The decision has to be made before the current subtask/maneuver ends. Also, decisions must be reevaluated if the state changes significantly.

- **Trajectory Level**
  The actual trajectory for executing the chosen driving maneuver is planned on this level with respect to goals specific to this level such as time-optimal trajectories or safety measures, e. g. keeping safety distances to obstacles.

- **Action Level**
  On this level, the agent directly controls and interacts with the system/vehicle to achieve the planned trajectory and ultimately accomplish the chosen subtask/driving maneuver.

The outcome of higher levels are passed on to the next lower level as requirements. On the other hand, lower levels can communicate the success or failure of their work to higher levels. The goal-directed action of cooperation partners (see Section 2.2.3) on all levels is emphasized by considering specific goals for each level and potentially different goals for each cooperation partner. Furthermore, the goals are assumed to be time-invariant for the current processing and meaningful with respect to the given level. Although the individual goals of the cooperation partners may differ, the goals have to be consistent such that arising conflicts can be resolved within the cooperation. Each layer in the butterfly model explicitly allows for direct communication and cooperation between the human and the automation via suitable interfaces (indicated by dashed lines in Figure 2.5). These interfaces may not be part of the original system, i. e. the vehicle in the exemplary application. They can be part of an *extended system*, e. g. a touchpad as utilized in conduct-by-wire concepts [FBB+14]. Also in the context of the research for this thesis, three interfaces for cooperative decision interfaces were implemented and examined which were based on touchpads, joypads and various displays, see Sections 4.1.1, 5.3.1, and 5.2.1.

Furthermore, the explicit modeling of direct cooperation on higher levels enables a straightforward integration of high LOA into the model. Consider the case of e. g. highly autonomous driving in which the action and trajectory levels are fully automatized, i. e. steering wheel and pedals are not present and the driver is only able to interfere with the vehicle via a maneuver interface, cf. conduct-by-wire concept [FBB^{+}14]. Hence, both, driver and vehicle automation, may be enabled to cooperate by negotiating upcoming maneuvers. This form of application can be described by replacing the two lowest levels of the model with a fully automated component that is integrated in the system. However, if the LOA can be changed flexibly, i. e. steering wheel and pedals are still present, an adaptation of the LOA e. g. similar to Baltzer et al. [BAMF14] could be applied as well.

Comparison to Other Layer Models of Human-Machine Cooperation

The outer appearance of the butterfly model in terms of structure does not differ greatly from the existing layer models. However, the *task abstraction levels* (see Section 2.2.3) are adapted to the context of executing a general task. This implies a shift in terms of the perspective on abstraction itself from time-horizon-based (i. e. strategy, tactics) to task-action-based (i. e. task decomposition, decision making). Nevertheless, there remains an analogy between the *strategic, tactical, operational,* and *execution* levels and the levels of the butterfly models. Additionally, the elements of *decision making* and *action implementation* of the conventional *perception-action cycle* (see Section 2.2.3) can be considered to be integrated into the task abstraction levels, see the task level names in the butterfly model. Even though the *perception elements* of the perception-action cycle are considered, as stated above, they are only implicitly visualized via arrows in Figure 2.5 for more clarity. Like in most other layer models, learning and training aspects (see [Ras83] and Section 2.2.3) are disregarded for reasons of simplicity. In contrast to other layer models, the goal-directed action on all levels with respect to individual goals of the cooperation partner is highlighted in the butterfly model. Furthermore, the butterfly model highlights the aspect of potentially direct communication and cooperation on all levels and explicitly considers itself to be lean and taxonomic. This implies that each layer resembles a placeholder for more specific models which serve as design model for human-machine cooperation on the respective layer. These specific models focus on forms of interaction (see Section 2.2.4) and especially on the abilities to cooperate (cf. "know-how-to-cooperate" [PLD02]). They also form the basis for cooperative automation designs which can be validated.

To conclude, the butterfly model provides a taxonomy for emancipated human-machine cooperation from an engineering and implementation perspective. Furthermore, it is suitable to intuitively circumscribe the research of this thesis and relate it to existing work in the context of emancipated human-machine cooperation: motivated by the success of established approaches for emancipated human-machine cooperation on the trajectory and action level [Fla19, LHFH18, Ing21], the research reported in this thesis targets human-machine cooperation on the decision level. Hence, the remaining thesis applies the butterfly model for classifying human-machine cooperation and elaborates on the decision level by means of mathematical behavior models of human-machine cooperative decision making (see Chapter 3), which forms the basis of experimentally evaluated automation designs to cooperatively decide with humans and resolve conflicts (see Chapter 4 and 5).

The next section motivates the research on the decision level of human-machine cooperation in more detail and discusses existing research with the same focus, providing details about automation designs and experimental investigations on cooperative decision making. This discussion reveals the research gap in more detail and is followed by a corresponding statement of the contribution of this thesis.

2.3 Human-Machine Cooperation on Decision Level

The research presented in this thesis investigates human-machine cooperation on decision level for four major reasons:

1. Reviewing the state of research on human-machine cooperative decision making (see Section 2.3.2) reveals a research gap, especially in terms of emancipated human-machine cooperation.
2. Research on emancipated human-machine cooperation at lower task levels, especially at action level, has revealed a great potential for creating synergies and improving performance [Fla19, LHFH18, Ing21]. Hence, aiming to transfer this success to higher levels is a logical next step.
3. For an in general functioning human-machine cooperation, all task levels require consideration and attention when designing appropriate automated cooperation partners [PLI15, FAI+16]. Especially cooperative human-machine systems with the ability to resolve decision conflicts can be assumed to be more robust and flexible in application, therefore advancing their scope significantly compared to existing, tailored human-machine systems for narrow scopes.
4. Facing current trends of increasing capabilities of automation in cooperative human-machine systems, the conflict resolution among cooperation partners on the decision level is a current pressing issue [FPLV+20]: On the one hand, increasing capabilities of the automation will enable human-machine cooperation on higher levels and therefore the ability to resolve decision conflicts is a key feature of these future cooperative human-machine systems. On the other

hand, regarding the serious out-of-the-loop problems of human operators in cooperative human-machine systems with high LOA, the continuous involvement of humans on decision level has the potential to keep the human in the loop such that she or he is consequently able to properly supervise the automation on lower levels.

Furthermore, the research reported in this thesis *exclusively* focuses on the decision level. This is motivated twofoldly:

1. The LOA of newly developed systems increases constantly which allows for fully automated performance of cooperative human-machine systems on lower levels, i. e. trajectory and action level, see e. g. [FBB+14, ACM+18].
2. The development of cooperative human-machine systems taking into account all levels at once is highly complex. Therefore, focusing on one level at a time and integrating them in the end potentially reduces this complexity.

Therefore, the focus on decision level is suitable in the scope of this thesis which is one of the first investigations on (emancipated) human-machine cooperative decision making.

2.3.1 Definition and General Solution Approaches

Human-machine cooperation on decision level can be analogously defined to Definition 2.6 of human-machine cooperation with a slight refinement of the term *task* which is specified to *decision making*. This leads to the following definition.

Definition 2.7 (Human-Machine Cooperation on Decision Level)
Two cooperation partners, i. e. human and machine, are involved in cooperative decision making, i. e. in a cooperation on decision level, if they meet two minimal conditions.

1. *Each one strives towards decision making objectives and can interfere with the other in the cooperative decision making process.*
2. *Each one tries to manage the interference to facilitate the individual activities in the decision making process and/or the common task, i. e. reaching an agreement.*

The symmetric nature of this definition can be only partly satisfied.

Although the term *cooperative decision making* could be perceived with a broader scope, it is always associated with human-machine cooperation in this thesis. It origins from the term *decision making* that is associated with the reasoning of one agent in a decision scenario and which is extended to the cooperative case that requires *individual decision making* of potentially all cooperation partners involved and a communication *process* among them to reach an agreement, constituting the *cooperative decision making process*.

Again, note that in Definition 2.7 *interfere* has no negative connotation and that the abilities and authorities in *cooperative decision making* are generically defined and require further definition in case of a specific scenario. In the most extreme case in terms of ability, a cooperation partner could not be able to make decisions, e. g. due to lacking relevant information. In this case, this cooperation partner will probably leave the other partner to make the decision for them. However, in the general case considered in this thesis, it is assumed that all cooperation partners are able to some extent to take part in the cooperative decision making process.

Furthermore, the general Definition 2.7 of cooperative decision making yields a vast scope, e. g. human-robot collaboration, driving assistance systems, etc. In the exemplary context of highly automated driving, cooperative decision making may manifest itself as follows: the human driver and the automation individually evaluate maneuver options, individually decide for their maneuver preference and subsequently participate in a (communicative) process to reach a mutual agreement on one maneuver option which is eventually executed.

In what follows, the form of interaction between cooperation partners on decision level and based on that general solution approaches for the cooperative decision making challenge are discussed.

Interaction vs. Communication

When Norbert Wiener introduced the term "cybernetics" in 1948 to describe the relation of animals/humans and machines, he postulated that a successful cooperation requires some sort of communication [Wie61]. Considering this from a human-machine cooperation perspective and in contrast to cooperation on action level, cooperative decision making may rely on two communication channels with respect to the butterfly model, see Section 2.2.6: the direct, explicit communication channel on the decision level and/or the interaction channel via lower levels and a potential interaction system. Table 2.2 provides the typical features of these different channels in the context of human-machine cooperation [JMS+16, RIK+17].

In essence, the direct, explicit communication channel offers a more abstract, richer communication, if it exists at all, while the interaction channel may be perceived to be more intuitive but has a more limited information flow.

On this basis, three general solution approaches for the challenge of cooperative decision making can be defined.

Table 2.2: Features of available communication channels on decision level.

Features	Direct, Explicit Communication Channel	Interaction Channel
Symbols	abstract symbols	physical signals
Interpretation of Symbols	not required	required
Information Flow	high	low
Existence (typically)	artificial	natural

General Solution Approaches

The general solution approaches for human-machine cooperative decision making differ in their assumption on the authority distribution among the cooperation partners and the richness of the available communication channel(s):

- **Trivial Cooperative Decision Making due to Information Alignment**
  Given a direct, explicit communication channel with a fast and extensive flow of information (e. g. exchange based on stenography), the information basis for decision making of all cooperative partners can quickly be aligned. Furthermore, assuming all cooperation partners are able to equally process the information and reason about it to reach a mutual (higher) goal, then all cooperation partners develop the same preference and trivially agree. Hence, no (communication) process is required to reach the agreement, e. g. [GR86, SBP+18, JA19, TW19]. Note that in this setting the authority distribution is irrelevant. Apart from that, this setting is highly unlikely in the context of human-machine cooperation as decision scenarios are usually highly complex and the communication channels will not be as rich as required, especially considering human limited perception capabilities.

  In the simplified example of cooperatively determining a route to drive, the navigation system may be able to provide all relevant time information of all potential route options to the human driver who has no other information to add. In case both cooperation partners pursuit the mutual goal of minimizing travel time, both will decide trivially for the same route.

- **Leader-Follower Approach**
  In this approach, the authority among cooperation partners is unequally distributed, putting one cooperation partner in the lead and hence also avoiding an extensive communication process to reach an agreement, e. g. [MM95, BK17, TI17]. Therefore, this approach is suitable for situations in which cooperation partners communicate via the limited interaction channel. Beyond that, the cooperation partner with minor authority might have important insights for the

decision making but is continuously overruled. Therefore, this approach is unsuitable in situations in which all cooperation partners have legitimate interest to participate in the process of decision making, e. g. if they obtain different but equally valuable information about the decision scenario. One solution to this can be a dynamic leader-follower role assignment that relies on a deterministic, universally valid assessment of cooperation partners' performance with respect to decision making.

In highly automated vehicles, the vehicle could e. g. assess how distracted the human driver is at any point in time. As soon the driver is distracted, the vehicle's automation takes control authority of e. g. maneuver selection. It hands back the control authority to the driver whenever the distraction disappears, although the automation may have legitimate reasons to decide differently than the human, e. g. due to different information bases for making the maneuver decision.

- **Emancipated Cooperative Decision Making Process**
  This approach assumes equal authority among cooperation partners and allows for different abilities with respect to decision making, aiming for an improved cooperative performance, e. g. [OKSB12, VKG14, OGD17, CHS21]. It therefore has the potential to yield a solution that is mutually agreed on by all cooperation partners. However, there is a risk of not reaching an agreement if both cooperation partners are unyielding. In terms of communication channels, this approach usually does not require the extensive information flow of a direct communication channel. Nevertheless, in order to avoid misinterpretation of symbols, the direct communication channel may be preferable compared to communication via the interaction channel.

  In the context of highly automated vehicles, the vehicle's perception of future traffic is outperforming the human abilities due to car-to-x communication. The opposite could hold for the perception of the rapidly changing close by traffic situation. In this case and with both cooperation partners pursuing a minimal travel time, the emancipated combination of both abilities to perceive traffic could be beneficial in selecting appropriate driving maneuvers. However, the traffic assessment results in a lot of information which cannot easily be shared among the cooperation partners. Therefore, an emancipated cooperative decision making process has the potential ability to implicitly fuse the information and yield a driving maneuver both cooperation partners mutually agree on.

The following section discusses the state of research of human-machine cooperative decision making which typically considers either the leader-follower approach or rarely an emancipated cooperative decision making process.

2.3.2 State of Research

Like research on general human-machine cooperation, research on human-machine cooperative decision making can be observed as the consequence of considering firstly individual decision making of humans or automated agents and subsequently decision making in groups of multiple equal agents, i. e. groups of humans or groups of automated agents, see Figure 2.6. Therefore, this section briefly presents some exemplary research on decision making of individuals and within classes of equal agents. Thereafter follows the discussion of research on decision making in the context of human-machine systems in more detail.

Figure 2.6: Evolution of decision making: from individual decision making over decision making of multiple agents of the same class (e. g. humans or automated agents) to decision making in the context of human-machine cooperation.

Decision Making of Individuals and Within Groups of Equal Agents

Research investigating human individual decision making began in the middle of the last century in the context of economics: Researchers tried to mathematically describe, understand and predict human decision making behavior when facing economical benefits and risks, e. g. in gambling or buying insurances, leading to extensive theories of expected utility [FS48] that have been advanced up to the present [KT79, KR14]. Besides this (more or less) static economical context, biologists investigated human decision making in the dynamic domain of human motion to understand the decision making process in terms of selection, planning and controlling of goal-directed human movements, see review of Gallivan et al. [GCWF18]. Furthermore, engineers developed and validated threshold models of human decision making, e. g. in dynamic process control to understand and predict how a plant operator detects events and selects actions when supervising multiple process measurements [GR82, GAR86].

Apart from human individual decision making, automated agents required decision making capabilities in the course of increasing automation in the last century. Therefore, engineers developed various decision making strategies and integrated these into hierarchical automation designs [GDW91, BYK+02].

Naturally, engineers extended these individual decision making capabilities towards multi-agent systems to allow for decentralized decision making of distributed artificially intelligent systems (see overview by Millot and Mandiau [MM95]) by means of methods such as prioritization and auctions. Exemplary scopes were task allocation [JCH16] and the coordination of autonomous airplanes or vehicles [SVP11, DP14, TLL+18]. Also in the course of research for this thesis, a decentralized path planning approach for cooperating autonomous mobile units with conflict resolution by means of a prioritization approach was introduced [RPFH19a].

Apart from these rather application-specific solutions in terms of cooperative decision making in action or trajectory planning (see task levels in the butterfly model in Section 2.2.6), a prominent theory developed over the last decade to formalize conflict resolution among automated agents in a more abstract way is called *negotiation theory* [Baa16]. Corresponding models consist of agent models and an interaction protocol along which the agents have to communicate offers. Each agent model comprises a utility function for evaluating offers and decision options, an acceptance strategy determining when to accept an offer of another agent, and a bidding strategy for generating own offers. Negotiation theory offers various application examples, e. g. supply chain management [Fin04], service distribution [ZR89], and traffic management of automated vehicles [ASM+05, YHS07]. Furthermore, there are many bidding /conflict resolution strategies [RS06, HL14] available as well as identification approaches for agents' negotiation behaviors [CJ04, HT08, MHM11].

In the context of human group decision making, one also observes two bodies of literature differing slightly in their scope: One type of research is concerned with human-human interaction or cooperation by providing some experiments investigating cognitive and neural processes in human joint action (e. g. [SBK06]) and sensorimotor control in joint action and planned coordination (e. g. [BK17]). Also engineers experimentally investigated haptic human interaction and found that humans are able to communicate and negotiate simple intentions haptically: Reed et al. [RP08] and Groten et al. [GFKP13] investigated paired human subjects who had to track conflicting reference trajectories while facing haptically coupled input devices. Weel et al. [WSA+18] examined the motion control in conflict situations of couples walking hand in hand on a Christmas market. The same insights were obtained in the realistic setting of driving assistance in case the driving assistance system is simulated by a human [JMS+16]. Similarly, an experiment in the course of the research on this thesis also yielded the insight that humans are able to cooperatively decide: two subjects were haptically coupled by means of force-feedback steering wheels and faced a dynamic evasive driving scenario which created conflict situations. In general, the subjects were able to successfully and cooperatively resolve the conflict situations [RGFH18].

The other type of research in the context of human group decision making focuses on mathematical models of abstract decision scenarios, e. g. in the economical context [MCAV19]. The most noticeable research of this type is summarized by *game theory* [FT91]: It provides models and analysis of decision scenarios with multiple intelligent, selfish entities involved. These entities are typically humans or animals and are called players. As a result, this theory usually describes game settings and constraints as well as provides and/or analyzes solution concepts, e. g. equilibria. In the context of dynamic group decision making, game theory offers various models such as revision games [CE08, CKLS14], bargaining games [Rub82, AG00] and the war of attrition that models conceding behavior in a competition [May74, BC78, BK99]. To be more specific, the war of attrition describes the concession behavior of players in an incomplete information setting, i. e. the players are unaware of the detailed reasoning of the other players. Application examples are hierarchic encounters in animal populations [May74] and market competitions [BK99].

With this research background concerning decision making of individuals or in groups of equal agents, researchers started to investigate human-machine decision making scenarios and transferred and adapted some aspects of this previous research.

Decision Making in the Context of Human-Machine Systems

In general, human-machine systems are required to make decisions in increasingly complex fields of application. Aside from simple, static authority assignments in terms decision making, the ability to cooperatively decide and resolve naturally occurring conflicts among cooperation partners is considered a key feature of automation designs in successful and robust cooperative human-machine systems aiming for a large area of applications [FPLV$^+$20]. Therefore, researchers developed approaches enabling the machine to actively participate in cooperative decision making. For further discussions, all resulting and existing approaches can be categorized by the authority allocated to the machine in cooperative decision making:

1) **Leader-Follower Paradigm**
   The authority in cooperative systems designed according to this paradigm is assigned to the leader who is in most cases the human. The follower may propose the own preference to the leader but only if the leader is absent the follower is able to enforce this preference. Therefore, in terms of authority assignment, designs obeying this paradigm are plain and well-defined. Apart from that, this paradigm is applied for various reasons such as liability and human acceptance.

2) **Decision Support Systems**
   The automation proposes decision options and a potential preference to the human who is in the lead and makes the decision.

3) **Dynamic Authority Assignment**
   The authority of the automation with respect to decision making within the cooperation is dynamically assigned considering the congruence of decision between human and automation, i. e. the follower-role of the automation is

dynamically shifted to the leader role if the human (potentially implicitly) accepts decisions of the automation. Ultimately, the human stays in the lead as the automation gives in if its decision is opposed by the human.

4) **Equal Authority Assignment**
Both human and automation engage in cooperative decision making as equal partners.

Approaches of the first and third categories implement the leader-follower approach for human-machine cooperative decision making introduced in Section 2.3.1 while the approaches subsumed in the second category try to trivialize cooperative decision making by providing proposals or other information. Approaches of the last category aim for or investigate emancipated cooperative decision making processes. In the following the existing approaches are discussed in more detail along these five categories.

The majority of applications in the context of human-machine systems employs the leader-follower paradigm with the human as the leader [CD13, JSB13, BK17, TI17] (rarely the automation is in the lead, e. g. [MM95]) or considers cases in which the task can be split in complementary but divisible subtasks such that the human and the machine work in parallel but not together on one subtask [JSB13]. The three major reasons for this are simplicity (automation design must not consider human behavior [BK17]), liability (if in the lead, the human clearly stays responsible for the decisions of the entire human-machine system [FPLV$^+$20]) and human comfort/acceptance (potentially disruptive decision behavior of the automation is avoided, therefore approaches aim for reducing conflicts to zero [GR86, SBP$^+$18, JA19, TW19]).

Closely related are decision support systems which aim at supporting the decision making of the human leader: Hindriks and Jonker [HJ09] addressed the potential mental overload of humans facing a complex decision situation with many options, aspects and stakeholders. To offer support to the human in these situations, they proposed a concept and architecture for a "pocket negotiator" which has to be provided with a description of the decision scenario and displays useful hints during the negotiation. Similarly, Suehiro et al. [SWS19] proposed a driving assistance system for decisions in lane merging to reduce cognitive load of drivers. The system is based on a human decision making model of drivers choosing merging positions. By means of this model, the systems predicts the merging gap and proposes the corresponding velocity to the driver. The corresponding experiment indicates reduced cognitive load and difficulty in decision making for the driver. Also in the field of driving assistance systems, Della Penna et al. [DvA$^+$10] designed an assistance system which reduces steering wheel stiffness to encourage faster decision making of drivers facing several evasive maneuver options. The authors emphasized that the decision capability and authority should stay with the driver but should be supported. Therefore, the driver is able to compensate the reduced stiffness. The experimental results show less crashes, decreased response times and control effort. To solve conflicts in cooperative control of highly automated vehicles, Baltzer et al. [BAMF14] introduced the concept of "arbitration": For controlling a highly automated vehicle, the driver and the assistance system interact via haptic multi-modal interfaces to navigate/guide/ control the highly automated vehicle. Via specific "interaction mediators" for the different driving task levels, the assistance system proposes a suitable action to the driver who can intercept or (implicitly) approve the action before it is potentially executed. In cases of emergency, the driver is firstly warned and ultimately "decoupled" from the driving task such that the automation solely executes actions to reach a safe state. Experiments proved the effectiveness of the concept. Upon this, the "conduct-by-wire" principle [GHW+11, FBB+14, FKGH15] was introduced for highly automated vehicles which do not require a manual stabilization of the vehicle and are guided by means of maneuver commands. To this end, maneuver interfaces have been developed to present maneuver options, indicate the preferred option of the automation and perceive the selection of the driver. The interfaces range from touchscreens and head-up-displays [KSB10, KFS+12, FKB+12, FKBG12] to driver gesture recognition [FDM+20]. By means of the driver's ability to decide for the maneuvers or supervise the maneuver decisions of the automation, the driver is kept in the loop and experimental evaluation reveals increased cooperative performance, reduced human workload and increased driver acceptance [FBB+14, FKGH15]. Walch et al. [WSH+16, WWM+19] also considered a highly automated vehicle which can be guided on a maneuver basis. The vehicle offers potential future maneuvers and the driver is able to approve the default option or select another one via a touchpad. Participants in the corresponding experiment reported a high usability and satisfaction with the proposed form of vehicle interaction. Motivated by the same area of application, Weßel et al. [WAS+19] proposed the concept of "self-determined nudging" which tries to support humans by nudging to make decision according to values and in situations the human authorized. Pacaux-Lemoine et al. [PLHSC20] proposed a decision support system in the context of a teleoperated robot: The robot is controlled by a human operator via an "emulated haptic feedback" brain-computer interface for selecting the direction the robot is driving (i. e. left, right, straight). To avoid obstacles, the automation increases the mental effort required to steer towards detected obstacles. In contrast to the above discussed decision support systems, the ultimate decision in which direction to drive is made by the automation to account for the low speed of the interface and hence potentially greatly delayed detection of human (thought) inputs. A conducted study showed the benefits of the emulated haptic feedback compared to operating the robot without it.

Another prominent category of research is concerned with dynamic authority assignment: Fern et al. [FNJT07] developed an assistant partially observable Markov decision process (POMDP) to observe a goal-directed behavior of a human, estimate the human's goal and decide on assistive actions. These action selections were customized to the individual users. The concept was evaluated in simulated environments with human subjects and showed substantial reduction of human effort. In the context of robotics, Kheddar [Khe11] proposed a control concept for humanoid service robots with the purpose of a dynamic leader-follower assignment based on the concurrence of the human's and the robot's motion goals. The aspiration was the development of a robot which is either passive or proactive if the motion goals are similar. However, no implementation details or results were reported. In contrast to this, Thobbi et al. [TGS11] published the results of an actual experiment in which a robot and a human were supposed to jointly lift a table. The robot was equipped with two controllers: One was reactive to the human motion, the other was proactive as it took the prediction of human motion into account. The switching of the controllers and hence of the authority distribution was influenced by the robot's confidence in the prediction of human motion. The experiment yielded improved cooperative performance.

Similarly, two collaborating research groups [MLK+12] developed a dynamic "role" assignment method for a human-robot team with the aim to assist but not disturb the human in joint task executions. The first implementations of the dynamic authority assignment were based on the alignment of human's and robot's forces on a joint work piece. The robot gradually increased force contribution if forces were aligned and reduced its contribution if this was not the case. Corresponding experiments revealed objective benefits in cooperative performance. However, participants perceived the force adaptation process as not transparent [OKSB10, MLK+12]. The authority assignment strategy was then advanced to an adaptation depending on human intention recognition in haptic collaborations with similar experimental results [KSB13]. Upon this, the intention recognition was refined by a data-driven stochastic model of human motion behavior. Additionally, the authority assignment was also advanced to allow for recessive to dominant attitudes of the robot depending on the uncertainty of human motion modeling and potential risk of the joint action. An experiment proved the increased helpfulness of the assistive system and human effort minimization [MLH15]. Corredor et al. [CSP14] developed an authority assignment strategy for teleoperation assistance with the aim to leave the human operator in the lead. To this end, the assistive force was dynamically adjusted depending on the concurrence of forces to track a reference trajectory.

An example of the dynamic authority assignment in driving assistance of highly automated ground vehicles is the "H-mode" introduced by Altendorf et al. [ABH+16, ABC+16]. It is inspired by the "(H)orse" metaphor of Flemisch et al. [FAC+03] from the same research group: the interaction of a rider and horse served as a metaphor for the development of assistance systems and their interaction with the driver. In the H-mode approach, the driver is supported by the assistance systems with various levels of automation. The change of the levels of automation is mainly initiated by the driver either by tight grasp of the steering wheel (reducing the degree of automation and resembling holding the reins tightly in the H-metaphor) or by pushing a button for in- or decreasing the degree of automation. The assistance system only initiates switches of automation degree in emergency situations [ABH+16, ABC+16].

The so far discussed models are strongly human-centered, i. e. they are mostly designed with an assistive objective. Furthermore, no mathematical behavior model of human decision making is utilized and the authority among the cooperation partners is not equal. However, there is also some research on equal authority assignment between human and automation. Vahidov et al. [VKG14] investigated bilateral and multi-bilateral negotiations between automated agents and humans in electronic marketplaces. The authors established a model of this scenario by means of negotiation theory and experimentally evaluated the performance of agents. However, the authors did not model, identify or adapt to human negotiation behavior. In another example, Oguz et al. [OKSB12] established a haptic cooperation game: one human and one automated player are haptically coupled and earn rewards depending on their cooperative (or selfish) action. The scenario was modeled by means of a multi-stage static game and three automation behaviors[11] were experimentally evaluated. Although human and automated player possessed equal authority, the research is concerned with a series of static decision making scenarios and the corresponding decision making history rather than with the dynamic process of cooperative decision making to reach an agreement within one scenario. Also in the case of so-called *mixed initiative systems* in which a (usually mobile) robot is operated with different LOA such as teleoperation or autonomy, the human operator and the robot's automation possess equal authority to *initiate* a LOA switch. For such systems, Owan et al. [OGD17] developed a so called *mixed-initiative control switcher*. In case the agents disagree whether or not to change the LOA, the robot's automation will drop its initiative for or resistance against the LOA change according to fixed time thresholds, ultimately giving the lead to the human. Chiou et al. [CHS21] propose another mixed-initiative control switcher focusing on when the robot's automation is taking the initiative to switch the LOA. They apply fuzzy control methods and adapt parameters to human (i. e. *expert*) behavior which results in the *expert-guided mixed initiative control switcher* (EMICS). Although EMICS is less intrusive than its predecessors, it may still lead to continual LOA switching, showing that the underlying conflicts for control of the robot between the human operator and the robot's automation is not resolved.

To summarize, Table 2.3 visualizes the major categories of research on cooperative decision making in the context of human-machine cooperation and the categories' key aspects. Note that for most approaches the category of decision support systems can be seen as a sub-category of human-in-lead due to its similarities in terms of automation authority. Besides this, Table 2.3 is the basis of the research gap discussion in the following.

[11] The three automation behaviors were either conceding relatively fast, conceding relatively late, and mirroring the behavior of the cooperation partner, denoted by *competitive*, *concessive*, and *tit-for-tat*, respectively.

Table 2.3: State of research on cooperative decision making in human-machine systems. The categories' key aspects are the following: The *automation authority* may range from follower to leader or may be equal to the human authority. The considered cooperative *decision making process* is either trivial, i. e. the agreement is found instantaneous, or only partially elaborated on in the respective work. The *human decision making behavior* may be modeled within some approaches and utilized in the automation design to avoid conflicts.

Categories	Range of Automation Authority	Decision Making Process	Human Decision Behavior
Human-in-Lead [GR86, TI17, SBP+18, TW19]	follower	trivial	rarely modeled and utilized
Decision Support Systems [HJ09, DvA+10, BAMF14, FBB+14] [SWS19, WWM+19, PLHSC20]	follower or leader	partially elaborated	rarely modeled and utilized
Dynamic Authority Assignment [FNJT07, OKSB10, Khe11, TGS11] [MLK+12, MLH15, ABH+16]	follower to leader	trivial	rarely modeled
Equal Authority Assignment [OKSB12, VKG14, OGD17, CHS21]	equal	partially elaborated	not modeled

2.4 Research Gap, Questions and Contributions

Regarding the state of research, the gap in research can be formulated upon which the research questions addressed in this thesis are stated.

Research Gap

The above summary of research on cooperative decision making in the context of human-machine systems provided in Table 2.3 reveals many approaches that deal with cooperative decision making to some extent. Within the categories human-in-lead, decision support systems and dynamic authority assignment of Table 2.3, most research tries to avoid conflicts between human and machine, either by means of intent recognition (decision support systems) and/or by (implicitly/ultimately) giving the human the leading role in the cooperation (human-in-lead, dynamic authority assignment). The minority of approaches deals with equal authority assignment between both agents, i. e. with emancipated agents. Although some approaches within this group consider some of the following aspects, there is no approach that

- considers the case of emancipated cooperative decision making between human and machine and
- properly focuses on the decision level of human-machine cooperation by
  - allowing for non-trivial cooperative decision making processes which lead to mutual agreements and by
  - utilizing suitable mathematical models of human decision making behavior, especially with focus on modeling human concession behavior in cooperative decision making.

However, enabling machines to take part in emancipated cooperative decision making processes with a human and to adopt human-like strategies may yield synergies and high user acceptance even in conflict situations: perceiving the automation like an emancipated cooperation partner, i. e. like another human, the conflict resolution may be as successful as research has revealed for the conflict resolution between two humans, see [RP08, GFKP13, JMS+16, RIK+17]. Furthermore, one cooperation partner's reasons for an initial decision which caused the conflict cannot simply be ignored by the other cooperation partner. As an example, the driver of a highly automated vehicle could not just ignore the decision of the vehicle's automation and the corresponding reasons for avoiding an unfavorable situation.

In order to investigate emancipated human-machine cooperation on decision level, a suitable automation design for the machine is required. To this end, this work utilizes a consistent model-based design approach. This approach offers several advantages compared to a heuristical design approach: It allows to introduce existing white-box knowledge of the considered human-machine cooperation on decision level. Additionally, it enables a comprehensible, explanatory description of the cooperation and of the automation behavior. Given this knowledge and description, a mathematical behavior model and hence an automation behavior similar to the respective mental model of humans may be generated which potentially leads to high user acceptance, see Section 2.2.2. Furthermore, the model-based design approach allows for a compartmentalized validation process and a replicable and easily adjustable design of the automation in new areas of application. Following this model-based approach to establish a suitable automation design, adequate mathematical behavior models of human-machine cooperative decision making are required. To eventually reveal potential benefits of the emancipated human-machine cooperation and although some research experimentally investigated aspects of cooperative decision making, the new models and automation designs demand for an innovative experimental design due to their exclusive focus on the decision level of human-machine cooperation.

In consequence of these research opportunities, this thesis addresses the following research questions and provides associated contributions.

Research Question 1

How to explicitly model cooperative decision making regarding human and machine as equal partners and considering human abilities as well as human behavior in a cooperative decision making process?

Contribution 1

Following the human-machine cooperation modeling approach via emancipated cooperation partners (dashed arrows in Figure 2.2), a first *meta-model of human-machine cooperative decision making* is introduced including a set of requirements resulting from human participation. Based upon this model design template, two novel mathematical behavior models for human-machine cooperative decision making are proposed: the *adaptive negotiation model* with its origin in negotiation theory and the *n-stage war of attrition* advancing game-theoretic models, see Section 2.3.2. Both treat the cooperation partners as equal in terms of authority and ability. Furthermore, the cooperation partners are modeled to individually evaluate and decide on decision options and mutually agree in a *process of cooperative decision making*. Additionally, human behavior in cooperative decision making is explicitly considered in both models to increase user acceptance. In the case of the adaptive negotiation model, this includes the *identification* and the *adaptation* towards the identified individual human behavior in the course of cooperative decision making. Moreover, a theoretical statement for the adaptive negotiation model is derived providing a *guarantee for finding an agreement* and hence for successfully resolving conflicts in cooperative decision making. In the case of the n-stage war of attrition, it is shown that the proposed game-theoretic strategies lead to a *perfect Bayesian equilibrium*. An overview and the relation of the models presented in this thesis is provided in Figure 2.7.

Research Question 2

How to design an automation based on a mathematical behavior model of cooperative decision making which is capable of participating in an emancipated cooperative decision making process with a human?

Contribution 2

After the introduction of the two mathematical behavior models of human-machine cooperative decision making, i. e. the adaptive negotiation model and the n-stage war of attrition, the models' suitability for describing human decision making behavior, more precisely human concession behavior in cooperative decision making processes, is investigated: the results of a corresponding study are presented

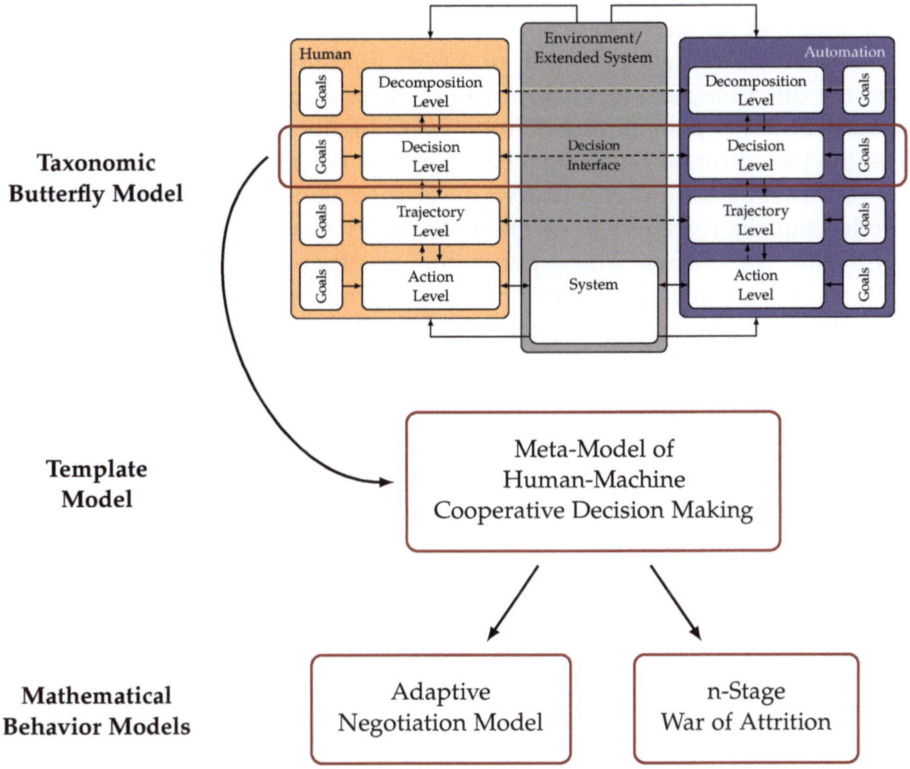

Figure 2.7: Overview and relation of the models presented in this thesis.

which prove the *models' suitability*. To complete a first holistic framework for human-machine cooperation on decision level, *automation designs for both proposed mathematical behavior models of human-machine cooperative decision making* are introduced. Furthermore, general guidelines for the implementation of an automation capable to participate in human-machine cooperative decision making are provided.

Research Question 3

Are there benefits of applying automation designs based on human-machine cooperative decision models (see Research Questions 1 and 2) to human-machine cooperation on decision level compared to state-of-the-art approaches?

Contribution 3

At first, a *general experimental evaluation approach* for investigating human-machine cooperative decision making is introduced due to missing experiments which exclusively focus on the decision level of human-machine cooperation. The approach comprises a set of guidelines and appropriate measures for suitable experimental designs. On this basis, two *experiments* are presented regarding two different application domains: teleoperated mobile robots and highly automated driving. In these settings, the two automation designs based on the proposed mathematical behavior models of human-machine cooperative decision making, i. e. the adaptive negotiation model and the n-stage war of attrition, are *experimentally compared* to relevant state-of-the-art approaches. The experimental results provide *first empirical evidence that the new automation designs significantly outperform the state-of-the-art approaches* in terms of objective cooperative performance. Similarly, the subjective evaluation results reveal a preference of the new automation designs.

The remaining thesis strives to answer these research questions and to fill the corresponding research gap by elaborating on the contributions. As a first step, the next chapter introduces a meta-model and the two mathematical behavior models of human-machine cooperative decision making.

3 Models of Human-Machine Cooperative Decision Making

In this chapter, a new theory on cooperative decision making in the context of human-machine cooperation is proposed to answer the first research question elaborated in the previous chapter: At first, a *meta-model of human-machine cooperative decision making* is proposed in Section 3.1 due to missing previous work on human-machine cooperation with model-based automation designs for the decision level. The meta-model describes the key properties of a cooperative decision making process and takes into account the requirements resulting from human participation. Applying the meta-model as a design template of the human-machine cooperation on decision level (see models' overview in Figure 2.7), two mathematical behavior models of cooperative decision making are introduced: the *adaptive negotiation model* in Section 3.2 and the *n-stage war of attrition* in Section 3.3, which originate from negotiation theory and game theory, respectively. Although, both mathematical behavior models describe a cooperative decision making process and are adapted to human behavior, the models differ in some aspects such as the consideration of decision making deadlines and the mathematical modeling of the concession behavior of the cooperation partners.

3.1 Meta-Model of Cooperative Decision Making

In the following, a first meta-model of general cooperative decision making is introduced: It comprises the general setting description of cooperative decision making scenarios and the interaction mode of the cooperation partners in these scenarios. Furthermore, a set of requirements arising from human involvement and modeling limitations are given to delimit the mathematical models considered in this thesis. By means of these requirements and limitations the general meta-model definition is refined to the meta-model definition of *human-machine* cooperative decision making.

Due to the lack of preliminary work that investigated a model-based approach for cooperative decision making with human participation, the following requirements on and definition of the meta-model are based on own observations and thoughts in addition to isolated hints in literature.

3.1.1 Introduction to the Meta-Model

When observing cooperative decision making in a social context, e. g. humans bargaining [Rub82] or negotiating contracts [KW89], it becomes apparent that elements of the underlying process can be generalized: At least two decision makers, e. g. merchants, face a set of at least two decision options, e. g. price levels. The decision makers are individually able to evaluate the options with respect to their payoff, e. g. profit margin, and decide for one preferred decision option. However, due to the cooperative setting, the decision makers have to choose one mutually-agreed decision option. Therefore, they have to advance from individual decision making to a coordination process. Within this process, the decision makers, i. e. the cooperation partners, communicate by means of acting, e. g. offering price proposals, and observing the others' actions via a corresponding communication channel. The communication may be based on natural language or other forms symbolic signaling, e. g. electronic bits in stock trading. Event-based communication is the most generalized form in terms of timing and has to be typically assumed if humans are involved and no other interaction protocol is in place. Furthermore, a pressure for reaching an agreement is usually present [Rub82], e. g. due to approaching the market place closing time. Therefore, rational cooperation partners interact strategically [CHC04] such that an agreement is reached while maximizing the individual payoffs as much as possible.

These general observations can be transferred into the technology context regarding machine-machine and human-machine cooperation. As a consequence, the following meta-model definition formalizes this generalized description of a cooperative decision making process for the first time and comprises the involved entities, the setting they are in and the mode of their interaction.

Definition 3.1 (The Meta-Model of Cooperative Decision Making)
The meta-model of cooperative decision making comprises the following elements:

- *A set of cooperation partners* $P = \{1, \ldots, N\}, N \geq 2$.
- *A set of decision options* $D, |D| \geq 2$.
- *An event-based interaction model consisting of the following elements:*
  - *A set of actions* $A := \bigcup_{i=1,\ldots,N} A_i$ *where* A_i, $i \in P$, *describes the set of actions of cooperation partner* i. *Each action "a" implies the choice of a decision option* ($a \implies d$, $a \in A$, $d \in D$).
  - *A set of possible events* $E := \bigcup_{i=1,\ldots,N} E_i$ *where* E_i, $i \in P$, *describes the set of events which cooperation partner* i *is capable to perceive.*
  - *The system dynamics* $\mathcal{S}$ *which transform every action into an event and/or trigger events according to internal system states.*
- *A potential pressure for reaching an agreement.*

If the cooperation partners act rational, they possess the following abilities:

- *A cooperation partner* $i \in P$ *acts according to a strategy* $\sigma_i \in \Psi_i$ *which is defined as a mapping of a sequence of event-time-tuples* $((e,t)_k)_{k \in \mathbb{N}^+}$ *to a sequence of action-time-tuples* $((a,t)_l)_{l \in \mathbb{N}^+}$:

$$\sigma_i : \{((e,t)_k)_{k \in \mathbb{N}^+}\} \mapsto \{((a,t)_l)_{l \in \mathbb{N}^+}\}$$

  with $e \in E_i$, $a \in A_i$ *and* $t \in \mathbb{R}^+$. *The set of strategy sets of all cooperation partners is denoted by* $\Psi := \{\Psi_i \mid i \in P\}$.

- *A cooperation partner* $i \in P$ *is able to evaluate strategies by means of a payoff function* π_i *which assigns a payoff to each sequence of event-time-tuples resulting from a strategy combination of all cooperation partners:*

$$\pi_i\big(((e,t)_k)_{k \in \mathbb{N}^+} \big| (\sigma_1, \ldots, \sigma_N)\big) \in \mathbb{R}.$$

  A rational cooperation partner chooses a strategy which maximizes the individual payoff.

Note. *A cooperative decision making process is fully described by the corresponding sequence of events* $((e,t)_k)_{k \in \mathbb{N}^+}$ *with each action being transformed into an event by the system dynamics* $\mathcal{S}$.

The definition of events E, actions A and the system dynamics $\mathcal{S}$ primarily resembles a general model of a communication and interaction channel among cooperation partners. The definition of strategies and payoffs provides some general guidelines for the rational goal-oriented reasoning of the cooperation partners: The objective of each cooperation partner is provided by maximizing the payoff function while taking into account the course of the cooperative decision making process which results from the own decision making strategy and the ones of other cooperation partners. Furthermore, the pressure for reaching an agreement can be modeled by means of a disagreement sensitive influence on the payoff functions and/or on the system dynamics. The strategy can be seen as the general road map in a cooperative decision making process in which participants strive towards their objective of maximizing their payoffs.

Definition 3.1 provides some template elements for cooperative decision making models but does not consider human abilities. What follows is therefore the discussion of requirements human participation poses on models of human-machine cooperative decision making.

3.1.2 Requirements Due to Human Participation

The participation of humans in a cooperative decision making scenario implies the following requirements which constrain some aspects of the meta-model of Definition 3.1.

Human Form of Interaction

Without enforcing any interaction constraints, human interaction is based on *discrete events* at *undefined times* with a *limited interaction rate* [MG17], i. e. the interaction rate is rather low in comparison to the one of technical communication systems.

The key element of this requirement, i. e. the event-based interaction, is already included in Definition 3.1. Besides this, the interaction rate is greatly influenced by the numbers of decision options and actions available, e. g. small numbers are assumed to cause a rather low rate of interaction as there is less to explore. First and foremost small numbers of decision options and actions enable the human to comprehend a decision scenario. A reasonable number may be four decision options/actions due to the fact that the human "focus of attention at one time [has four as a] capacity limit" [Cow01]. In terms of the human mental short-term storage capacity slightly higher numbers are discussed in literature [Cow01]. Since these cognitive limitations of humans must be considered by the model of human-machine cooperation, the following assumption on the number of decision options is posed in a generalized manner.

Assumption 3.1. *The sets of decision options D, events E and actions A have a size which is sufficiently small such that the cognitive abilities of humans are not exceeded.*

State of Knowledge

To the knowledge of the author, there is no model of general human reasoning in a cooperative decision making process. Furthermore, it is in general not easy to transfer this reasoning from human to machine and potentially infer vice versa. Hence, models of human-machine cooperative decision making should consider an incomplete information setting, i. e. the source of reasoning and the reasoning process of humans is in general not explicitly available to other cooperation partners. For reasons of symmetry, this is assumed for all cooperation partners.

Assumption 3.2. *Cooperation partner models have to assume that they possess incomplete information about the other cooperation partners.*

Note. *The lack of information of the cooperation partners on other partners is not a hindrance when implementing cooperative decision making with human participation. In fact, an experiment conducted in the course of the work on this thesis found that two humans are able to cooperatively decide in a scenario in which only a limited haptic communication channel is available [RIK$^+$17].*

Human Rationality and Strategy Determination

Definition 3.1 comprises a general description of strategies of rational cooperation partners. Rationality describes the *depth of strategic thinking* in pursuing the objective, i. e. a (fully) *rational* cooperation partner strives to determine a strategy that maximizes the individual payoff in complete information settings or expected payoff in incomplete information settings whereas a *non-rational* cooperation partner acts randomly [Str14].

Humans exhibit a behavior of *bounded rationality* [Har17], i. e. they will maximize their payoff based on a finite cognitive level, described by the *cognitive hierarchy theory* [Nag95] and its enhancements to different scopes [CHC04, CHC16, AY21]. This is due to the fact, that humans do not possess unlimited cognitive power to assess their actions' impact without loss of time or other resources. For example, they are not generally able to assess the infinite circle of impact of their actions on the other cooperation partners' actions, on their actions, and so forth. Instead, they may stop after a specific *depth of thought*: In *level 0*, actions are chosen randomly; in *level 1*, the player chooses actions assuming all other players are of level 0; and so on [Nag95]. It is in general difficult to determine the level of rationality of a human. However, some experimental evidence indicates rather low numbers, i. e. level-1, level-2, or at most level-3 [CHC04, CGC06].

Other research on human decision making in rather simple, non-cooperative decision scenarios was able to fit *rule-based* models to human decision making actions and to utilize these models to predict human decision making actions, e. g. [GR82]. In more complex decision scenarios, hints for human *reflection and adaptation techniques* were observed [VKG14, GCWF18]. Examinations of human decision making behavior in the game theoretic context confirmed non-fully-rational human behavior and descriptive mathematical decision models with *probabilistic* influences could be fit to experimental human data [MP95, AGH04].

Given the above hints and observations in literature on the bounded rationality of humans, the suitability of behavioral models based on rules, reflection and adaptation techniques, and probability, and considering the regarded cooperative, incomplete information setting (see Definition 3.1 and Assumption 3.2), three general approaches for strategy determination in the context of this thesis are proposed:

- **Reaction**
  Cooperation partners react to events based on their own strategy without any reflection on the strategy of other cooperation partners while being in the cooperative decision making process. This approach is associated with a level-1 depth of thought.

- **Identification-Prediction-Action**
  Cooperation partners identify the other cooperation partners' strategies during the cooperative decision making process. On that basis, they are able to predict the consequence of their own choice of strategy and adapt it accordingly. Consequently, this approach comprises the reflection of decision making behavior and represents at least a level-2 depth of thought. However, with an increase of the depth of thought, strategy determination becomes more challenging and is no longer human-like [CHC04, CGC06].

- **Uncertainty-Action**
  Cooperation partners possess no detailed information on the other cooperation partners' strategy or payoff function. However, they have some probability information on the strategies or payoff functions which they utilize in their strategy determination. Hence, this approach also represents a level-2 depth of thought. As there is no more information available without utilizing some identification approach, this approach could be considered fully rational in the given information setting.

Each of these general approaches is rational to some extend and it depends on the mathematical behavior model of cooperative decision making which approach's application is suitable. These insights are summarized in the following assumption on models of cooperation partners in the context of human-machine cooperative decision making.

Assumption 3.3. *Cooperation partners are modeled with respect to bounded rationality and following one of the three general strategy determination approaches: reaction, identification-prediction-action, or uncertainty-action.*

3.1.3 Additional Assumptions and Limitations

Following the Definition 3.1 of the meta-model for cooperative decision making and the discussion of model requirements due to human participation in a cooperative decision making scenario (see Assumptions 3.1 to 3.3), a set of additional assumptions and limitations is introduced for reasons of models' manageability and applicability in the context of automation designs for human-machine cooperation on decision level.

The following assumption restricts the general decision scenario considered in this work for reasons of manageability: Decision options and communications symbols, i. e. events and actions, are limited to finite, discrete numbers which are known to all cooperation partners, allowing for straightforward interface designs and theoretical model analysis. For the same reasons, the form of interaction is set to be deterministic and time-invariant.

Assumption 3.4. *The general decision scenario is limited to:*

- *Discrete, finite sets of decision options D, events E and actions A which are identical for all cooperation partners, and*
- *A deterministic and time-invariant system $\mathcal{S}$, i. e. form of interaction of the cooperation partners.*

Due to human preference of interaction at undefined times [MG17], the timing of the interaction shall not be constrained to some potentially unintuitive communication protocol. Moreover, the presence of an element creating pressure to reach an agreement is required to make cooperative decision making worthwhile [SGC98]. In practice, this element may be e. g. a deadline $\mathcal{T}$ until which cooperation partners have to agree on one decision option [SGC98].

Assumption 3.5. *The timing of the interaction is unrestricted and an element creating pressure to reach an agreement is present.*

In the general decision scenario, this work considers two cooperation partners, i. e. the human and the machine (see Definition 2.1), with the following characteristics: For reasons of emancipation (see Definition 2.2), the cooperation partners' rights shall possess equal rights. To enable potential benefits of cooperative decision making, both partners shall be equally performant in terms of decision making, i. e. no cooperation partner is able to continuously outperform the other. However, cooperation partners may possess individual objectives and/or different individual information bases for decision making. For reasons of identification and reproducibility, the strategies of the cooperation partners in a cooperative decision making process shall be deterministic but may be based on probabilistic information. Furthermore, the considered strategies are limited to those which lead to a conceding behavior in the cooperative decision making process, i. e. cooperation partners strive towards an agreement on one decision option and cannot take back the proposal of a decision option. This limitation is introduced for reasons of manageability in the initial mathematical modeling and analysis of human-machine cooperation on decision level in this thesis. These characteristics of the cooperation partners are summarized in the following assumption.

Assumption 3.6. *Two cooperation partners, i. e. one human and one automated agent, are considered with:*

- *equal rights,*
- *equal performance in terms of individual rational decision making with different objectives and information bases,*
- *deterministic strategies which lead to*
- *a conceding behavior throughout the cooperative decision making process.*

With regard of training effects in human behavior described by Rasmussen [Ras83], this work focuses on first investigations of stationary human-machine cooperative decision making processes and neglects long-term learning for the sake of simplicity.

Assumption 3.7. *No long-term learning or training effects need to be modeled.*

3.1.4 Meta-Model of Human-Machine Cooperative Decision Making

The following definition summarizes all requirements for and assumptions on cooperative decision making models in the scope of this thesis.

Definition 3.2 (Meta-Model of Human-Machine Cooperative Decision Making)
The meta-model of human-machine cooperative decision making is the enhancement of cooperative decision making (Definition 3.1) by means of the requirements given by Assumptions 3.1 to 3.7. The key aspects are the following:

- *A discrete set of decision options D is available of a size which is sufficiently small such that the cognitive abilities of humans are not exceeded.*
- *Two cooperation partners* $P = \{\mathcal{A}, \mathcal{H}\}$*, i. e. the human* $\mathcal{H}$ *and the automation* $\mathcal{A}$*, are considered with*
  - *equal rights and abilities of individual decision making with bounded rationality and individual objectives,*
  - *incomplete information about the other cooperation partner,*
  - *deterministic and conceding strategies which are determined following one of the three general strategy determination approaches (reaction, identification-prediction-action, uncertainty-action).*
- *The cooperation partners interact on an event-basis. This interaction is deterministic and time-invariant. The sets of events E and actions A are discrete, sufficiently small and identical for both cooperation partners.*
- *A pressure for reaching an agreement has to be in place, e. g. a deadline* $\mathcal{T}$ *until which cooperation partners have to agree on one decision option.*

After the introduction of the meta-model of human-machine cooperative decision making and the assumptions on and limitations of models considered in this thesis, the following section explains the choice of two theories which serve as a basis to derive two mathematical behavior models of human-machine cooperative decision making in Sections 3.2 and 3.3.

3.1.5 Motivation for the Theoretical Basis of the Developed Models

In the discussion of the research gap with respect to human-machine cooperative decision making in Section 2.4, the two key aspects are the lack of approaches which consider a *non-trivial process* of cooperative decision making and the disregard of *equal authority* of the cooperation partners human and automation within this process. However, the state of research presented in Section 2.3.2 provides two prominent theories with models which incorporate non-trivial processes of cooperative decision making among emancipated cooperation partners: *negotiation theory* and *game theory*. Yet, negotiation theory usually only considers automated agents which can be programmed and game theory regards independent players such as humans which cannot be influenced by a system's designer. Facing the cooperation of human and machine, approaches and models of both theories cannot be applied directly. Nevertheless, there are some approaches with origins in either negotiation theory or game theory which were successfully investigated in some context of human-machine cooperation: E. g., Oguz et al. [OKSB12] examined human behavior in a series of static decision games without modeling the individual decision making process. Vahidov et al. [VKG14] investigate adaptive strategies in human-machine negotiation with a time-horizon of several days.

Consequently, the research reported in this thesis advances mathematical behavior models of negotiation theory and game theory to meet the requirements of the introduced meta-model of human-machine cooperative decision making (see Definition 3.2 and models' overview in Figure 2.7) and to close the gap between models of cooperating automated agents and models of cooperating independent players. The resulting models are the *adaptive negotiation model* and the *n-stage war of attrition* model. They differ in the general approach to model strategy determination, see Section 3.1.2: the adaptive negotiation model relies on the reaction or identification-prediction-action approach whereas the n-stage war of attrition utilizes the uncertainty-action approach.

Aside from a slightly different perspective of authority assignment and differing strategy determination approaches facing incomplete information scenarios, these models also close the gap between human-in-lead and automation-in-lead: customized models of negotiation theory comprise the urge to find mutual agreements between agents which will force agents to ultimately give in whereas the automation designs based on adapted game theory models focus on their independence and thus may not ultimately concede. However, in a practical application scenario, a final decision may be required at a fixed deadline. If cooperation partners cannot reach a mutual agreement before the deadline, this consequently leads to an ultimately higher authority of the human in case the automation is designed based on the adaptive negotiation model. The opposite holds for the application of the n-stage war of attrition. Therefore and despite all efforts, the state of equal authority in the context of human-machine cooperation will not be achieved if the cooperation partners cannot find a mutual agreement. However, this state is also not achievable in cooperation of automated agents nor in cooperation of humans for the same reason.

Figure 3.1 illustrates the relation of the leader-follower distributions and the developed models, i. e. the adaptive negotiation model and the n-stage war of attrition model. It thereby provides the motivation why this thesis elaborates on and investigates both models. The following two sections are devoted to the introduction of the two human-machine cooperative decision making models.

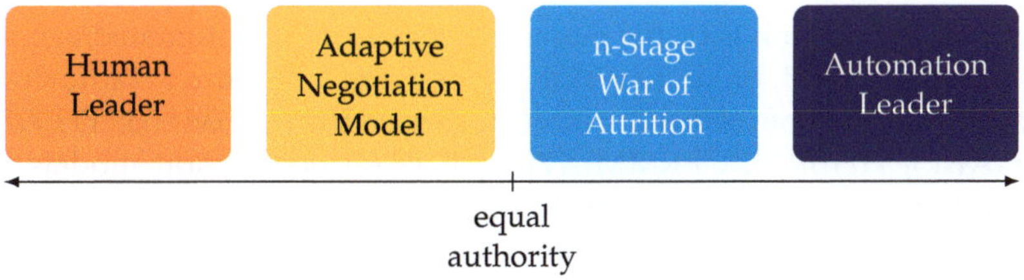

Figure 3.1: Relation of models based on negotiation theory and game theory to leader-follower models in terms of authority distribution.

3.2 Adaptive Negotiation Model

The following section introduces the adaptive negotiation model that enhances conventional negotiation theory by allowing for human-machine negotiations. This research was the result of two supervised master theses [Sch18, AW19] and led to two publications [RSFH19, RAFH20].

3.2.1 Introduction and Terminology

The following general statements about negotiation theory are derived from Baarslag [Baa16], one of the standard references in terms of negotiation theory. Negotiation theory originally provided models for multi-agent systems with *autonomous agents* to negotiate in conflict situations with potentially multiple *issues*. Within the negotiation process, i. e. process of cooperative decision making, the agents exchange *offers* representing decision options according to a *bidding strategy*. This strategy relies either on a *time-based concession strategy*, modeling negotiation pressure increasing with time, or on a *behavior-based concession strategy*, directly reacting to the other agents' negotiation behavior and actions, e. g. the tit-for-tat strategy. The latter type of strategy is prone to cause endless negotiations without any agreement. Agents accept or reject offers of other agents based on an *acceptance strategy* which is based on *utilities* the agents individually assign to these offers. For the case that no agreement is found until a certain deadline, it is common to define in advance a *conflict deal* all agents agree on. This is possible due to the fact that usually automated, i. e. programmable, agents are considered. The interaction of agents is defined by means of a *negotiation protocol*. In state-of-the-art negotiation models, *simultaneous* or *alternating* protocols are applied in which agents exchange offers simultaneously or in an alternating fashion, respectively.

In literature, many application examples of negotiation models for the design of negotiating autonomous agents are available. The scopes range from supply chain management [LC10, Fin04] to task and service distribution [ZR89, HSW05, KAL07] and buyer-seller scenarios in automated e-commerce [FSJ98, CW15, CJ04, WWY11].

Another area of application is traffic management in which automated agents within one domain, i. e. sea, land, or air, negotiate maneuvers to optimize traffic flow or evasion maneuvers in case of conflicting trajectories [WVI04, ASM$^+$05, YHS07, SVP11, DLDS13, GAB15, HHBR15, CPRMML17]. All these negotiation models were designed for automated, i. e. programmable, agents which communicate with a high rate and quantity. With regard to the targeted form of human-machine interaction and its limitations on communication among agents introduced in Section 3.1, these models are unsuitable for a direct adaptation to human-machine cooperation on decision level.

However, there are some approaches which consider human-machine interaction in the field of *human-agent negotiation*. Some models were used to implement negotiation support systems for humans, e. g. [HJ09]. Their aim is to support the human in multi-issue negotiations by providing suitable graphics which help to keep the overview of the negotiation. Furthermore, they try to compensate human negotiation errors due to impatience or emotion-driven actions. Vahidov et al. [VKG14] experimentally investigated human-machine negotiation in a buyer-seller scenario focusing on time-dependent and behavior-dependent bidding strategies which outperformed humans in negotiations. The results showed that in bilateral negotiations "competetive" bidding strategies are favorable but in general adaptive behavior strategies may yield benefits. However, such behavior adaptation requires information about the other agent's negotiation behavior. Human negotiation behavior has been found to be individual without the possibility to make general assumptions [OLK09]. Mell and Gratch [MG17] aimed at replicating human negotiation behavior by means of a web-based platform for multi-issue bargaining and by focusing on human features in the context of human negotiation participation: they paid great attention to the communication channel such that it allowed for low communication rate, speech and transfer of emotions. Furthermore, they allowed for irrationality and partial offer exchange in their automated agent designs. The negotiation setting was multi-issue negotiation in which agents have to iteratively negotiate a resource distribution. There was no eminent pressure for decision making and the automated agent only acted upon human offers or other communication events, resulting in an alternating offer negotiation protocol. The results showed that it is crucial to account for human capabilities in negotiations, especially in terms of communication. The authors advanced their research and negotiation models to account for more human-like traits such as making promises or to betray others [MLG20].

In summary, the few approaches in the context of human-agent negotiation do not entirely fit the modeling objectives of human-machine cooperative decision making of this thesis, see Section 3.1.4: they either only support or try to outperform humans in negotiations or replicate human negotiation behavior in situations with little pressure to reach an agreement. Despite this and the low number of human-agent negotiation models, the existence and success of these models encouraged the development of a negotiation model which suits the requirements of the meta-model of human-machine cooperative decision making: the objective of this new negotiation model is to represent a human-machine negotiation over a set of decision options D by exchanging offers $o \in O$ among two participating agents $i \in \{\mathcal{A}, \mathcal{H}\}$, i. e. automation and human. Although the general structure of conventional negotiation models can be inherited, i. e. utility functions, acceptance and time-based concession strategies, the introduction of the human into this automated agents' theory results in some design challenges for the components which can be derived from the introduced meta-model of human-machine cooperative decision making in Section 3.1.4. First and most importantly, the basis of reasoning is generally unknown. Hence, the exchange of offers is the only direct source of information for the automation. Second, but closely related, no conflict deal can be defined in advance. Third, the timely form of interaction among agents with human participation requires attention.

Consequently, the following introduction of the adaptive negotiation model focuses on the required enhancements of conventional negotiation models towards a human-compatible negotiation model. The specific enhancements are

- an *asynchronous negotiation protocol* to suit the discrete event character of human action and communication,
- the selection and application of an *identification* approach for identifying the other agent's behavior, i. e. an *opponent model*[12], that is able to perform on little information due to an expected limited rate of communication of the humans involved, as well as
- a generalized, explicit strategy for *adaptation* of negotiation tactics to draw advantage from deeper insights into the other agent's reasoning via the identification approach, and
- an *agreement guarantee* by means of the asynchronous negotiation protocol and a suitable concession design of the automated agent.

3.2.2 Model Definition and Overview

This section provides the definition of the *adaptive negotiation model* based on the requirements of human-machine cooperation on decision level and the model limitations considered in this thesis described in Section 3.1 altering state-of-the-art negotiation models as stated above.

---

[12] In the context of negotiation theory other agents are referred to as *opponents* which is then also the name origin of corresponding *opponent models* to identify their behavior. However, in this thesis' context of human-machine *cooperation*, the term opponent is avoided as agents are negotiating to reach an agreement and resolve conflicts.

Definition 3.3 (Adaptive Negotiation Model)
The setting of human-machine cooperative decision making for the adaptive negotiation model consists of the following components:

- *A set of two rational, adaptive agents $P := \{\mathcal{A}, \mathcal{H}\}$ (will be defined subsequently in Definition 3.4): $\mathcal{A}$ denotes the automation and $\mathcal{H}$ the human.*
- *A discrete, finite set of offers O which is identical for both agents. Each offer $o \in O$ is associated with one decision option d of a given discrete, finite set of decision options D but may be enriched by some additional information. In other words, an offer o is a communication symbol for which $o \implies d$ holds but not necessarily vice-versa. The notation o^t includes the time $t \in \mathbb{R}^+$ at which an offer is proposed.*
- *An asynchronous negotiation protocol [RSFH19], allowing the agents to communicate, i. e. exchange offers, on an event basis which suits human communication behavior. However, except the initial offers, simultaneous offers are prohibited, i. e. two offers o^{t_1} and o^{t_2} are required to be proposed at different times $t_1 \neq t_2$, $t_1, t_2 \in \mathbb{R}^+$.*
- *A negotiation deadline $\mathcal{T} \in \mathbb{R}^+$ until which agents have to agree on one decision option, i. e. one agent has to accept an offer of the other agent.*

In this model, a negotiation starts as soon as both agents (potentially simultaneously) placed initial offers. This point in time is defined as $t = 0$. In a conflict situation, i. e. agents favor different decision options, the rational agents concede by strategically proposing offers, which they cannot take back. Hence, agents establish a history set of offers $O_i^H \subset O$, $i \in \{\mathcal{A}, \mathcal{H}\}$. The negotiation ends when an agreement among agents is found.

Remark. *In the adaptive negotiation model's definition, the case of not reaching an agreement before the deadline is purposefully excluded. In conventional negotiation theory, this case is handled by the definition of a conflict deal which is impossible in the intended scope of human-machine cooperation with the requirement to consider both cooperation partners as equal. Hence, the following definitions and assumptions will provide a setting in which it is guaranteed that an agreement is reached before the deadline is met.*

The term *adaptive* in the name of the above defined negotiation model stems from the applied *adaptive agent model* which is defined in the following.

Definition 3.4 (Adaptive Agent Model)
The rational, adaptive agents $i \in P$ are modeled by means of the following aspects:

- *An individual, time-invariant utility function u_i, according to Definition 3.5 for evaluating offers $o \in O$, $|O| \geq 2$, i. e. $u_i(o) \in \mathbb{R}$.*
- *An acceptance strategy $\mathcal{C}_i$ (see Definition 3.6) determining whether to accept or decline offers o_j^κ (short for $o_j^{t_\kappa}$) of the other agent based on the offer's utility compared to the utility of an own current offer o_i^k at time t_k, taking into account the entire offer history sets O_j^H of the other agent, i. e.*

$$\mathcal{C}_i\left(u_i\left(o_i^k\right), \left\{u_i\left(o_j^\kappa\right)\right\}_{\forall o_j^\kappa \in O_j^H}\right) \in \{accept, decline\}.$$

- *A bidding strategy $\mathcal{B}_i$ for determining a (counter) offer o_i which is set to be a time-based concession strategy $\mathcal{E}_i$ (see Definition 3.8), modeling an increasingly concessive behavior over time t, i. e.*

$$\mathcal{E}_i(u_i, t) \in O,$$

  is utilized which is motivated by its successful application in the context of human-machine negotiation [VKG14] and by the presence of a deadline. Due to agents' rationality, agents will always propose offers in a sequence such that the offers' utilities strictly decrease, starting with their initial offer o^0 associated with the highest utility. This fact together with a time-invariant utility function explain why agents do not take back offers already proposed.
- *An identification module which agent $i \in P$ utilizes to identify the bidding strategy $\mathcal{B}_j$ and ultimately the concession strategy $\mathcal{E}_j$ of the other agent $j \in P$, $j \neq i$. The identification module is required to work on the expected limited communication and therefore little information exchange between agents (see Definition 3.9).*
- *An explicit, generalized adaptation module that allows agents to adapt their bidding strategy $\mathcal{B}_i$ based on the insights generated by the identification module regarding the other agent's negotiation behavior (see Definition 3.12).*

Any specific structure or parameterization of the agents' components introduced above is private information and remains unknown to the other agent.

Figure 3.2 provides an overview of the introduced adaptive negotiation model and the interaction between its components. Therefore, Figure 3.2 is a refinement of the block *adaptive negotiation model* in the models' overview depicted in Figure 2.7. Within the *basic negotiation model*, agents interact (i. e. communicate) according to the *negotiation protocol*, evaluate offers by means of an individual *utility function* and accept or generate offers via *acceptance* and *bidding strategies*. Through the *identification* module and the explicit *adaptation* component, agents are able to adapt their bidding strategy, i. e. negotiation behavior, with respect to the previously observed behavior of the other agent.

Furthermore, Figure 3.2 connects the components of the adaptive negotiation model with the aspects of the general identification-prediction-action paradigm, see strategy determination approaches in Section 3.1.2: after *identifying* the other agent's behavior, the adaptation module *predicts* the course of the negotiation and allows for an adaptation of the agent's bidding strategy, i. e. the agent's *action* determination. In broader terms, the action part of the model can be seen as the *tactics* of negotiation, leaving the prediction and adaptation part to resemble the negotiation *strategy*.

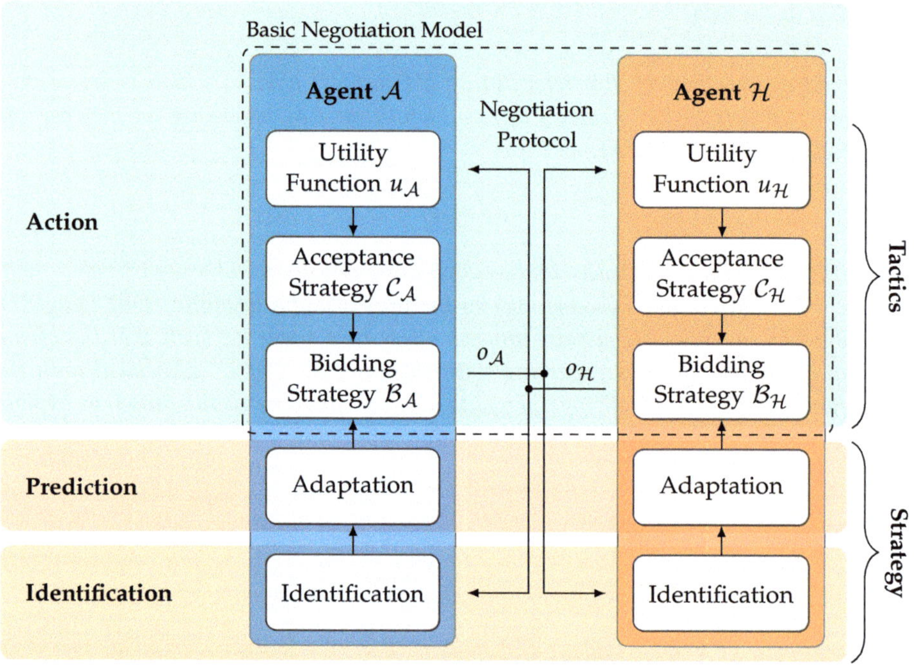

Figure 3.2: Overview of the adaptive negotiation model and its components' connection to negotiation strategy and tactics and to the aspects of the general identification-prediction-action paradigm. Agent $\mathcal{H}$ resembles the human and Agent $\mathcal{A}$ the automation.

3.2.3 Details of the Basic Negotiation Model

In the context of human-machine cooperative decision making, the basic negotiation model resembles the reaction part of the adaptive negotiation model (see strategy determination approaches in Section 3.1.1).

Figure 3.3 provides an overview of the reasoning and reaction process of one agent i in the basic negotiation model. In each cycle k of decision making, which corresponds

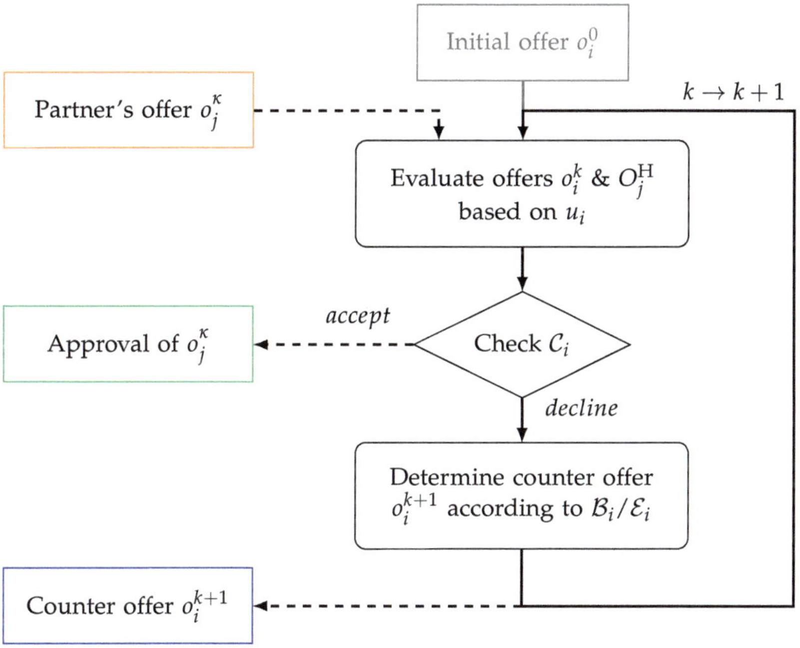

Figure 3.3: Overview of reasoning for one agent i in the basic negotiation model ($i, j \in \{\mathcal{A}, \mathcal{H}\}, j \neq i$).

to a time t_k, agent i evaluates its own current offer $o_i^k \equiv o_i^{t_k}$ and the offers of the offer history $o_j^{t_\kappa} \equiv o_j^\kappa \in O_j^{\mathrm{H}}$, $\kappa < k$, established by the cooperation partner, agent j, at earlier times t_κ by means of the utility function u_i. Then the agent decides whether the other agent's offer should be accepted or rejected according to its acceptance strategy $\mathcal{C}_i$. If the other agent's offer is declined, the agent determines a new counter offer o_i^{k+1} in line with the own bidding strategy $\mathcal{B}_i$, i. e. in this case the concession strategy $\mathcal{E}_i$. This offer is presented to the other agent. The next cycle may start at potentially any time unless an agreement or the deadline $\mathcal{T}$ has been reached.

Regarding its application in the context of human-machine cooperative decision making, the components of the basic negotiation model are defined in greater detail in the following.

Utility Function

In line with state of the art approaches and without loss of generality the proposed structure for the utility functions is a linear combination of normalized evaluation functions $\bar{b}(\cdot) \in [0,1]$ for various aspects of the negotiated issues. By means of normalized weights in the linear combination, this leads to a normalized utility function $u := \bar{u}$ and hence to comparable evaluations of different negotiation scenarios.

Definition 3.5 (Normalized Utility Function)
Based on normalized evaluation functions $\bar{b}_i(o) : O \mapsto [0,1]$, the normalized utility function $\bar{u}_i(o) : O \mapsto [0,1]$ of agent $i \in \{\mathcal{A}, \mathcal{H}\}$ is defined as their linear combination:

$$\bar{u}_i(o) := \sum_l w_{i,l} \cdot \bar{b}_{i,l}(o) \tag{3.1a}$$

with $\sum_l w_{i,l} = 1$.
Furthermore, the normalized utility function has to enable a meaningful differentiation of offers within each negotiation scenario, i. e. the offers' utilities have to be unique:

$$\bar{u}_i\left(o^1\right) \neq \bar{u}_i\left(o^2\right) \quad \forall o^1, o^2 \in O, o^1 \neq o^2. \tag{3.1b}$$

Note. *In general, negotiation theory allows for time-dependent utility functions, i. e. $\bar{u}_i(o,t) : O \times \mathbb{R} \mapsto [0,1]$ and $\bar{b}_i(o,t) : O \times \mathbb{R} \mapsto [0,1]$. Due to the requirements of the meta-model of human-machine cooperative decision making (see Definition 3.2), only time-invariant utility functions are considered in this thesis, see definitions of $\bar{u}_i(o)$ and $\bar{b}_i(o)$ in Definition 3.5.*

As an example for a time-invariant utility function, consider the use case of navigating a vehicle in which cooperation partners may negotiate over different routes before starting to drive. In this example, the routes represent the decision options. To evaluate each route, two normalized evaluation functions given by the fuel savings on a route relative to the maximum fuel savings of all routes and the travel time savings on a route relative to the maximum travel time savings of all routes could be used. The weighted sum of these evaluation functions constitute the utility function. The cooperation partners assigning different utility values to a given route and hence having different preferences can result from cooperation partners weighting fuel savings and time savings differently. Another reason for different utility values and preferences can be varying assessments of fuel costs or travel time in a given situation, e. g. due to different information bases.

Acceptance Strategy

Considering the concession behavior of both agents and their rationality due to which they cannot take back offers, the acceptance strategy for both agents is defined as follows.

Definition 3.6 (Acceptance Strategy)
Applying the Normalized Utility Function Definition 3.5, the acceptance strategy C_i of both agents $i \in \{\mathcal{A}, \mathcal{H}\}$ is set to:

$$C_i\left(\bar{u}_i\left(o_i^k\right), \left\{\bar{u}_i\left(o_j^\kappa\right)\right\}_{\forall o_j^\kappa \in O_j^H}\right) := \begin{cases} accept, & \exists\, o_j^\kappa \in O_j^H : \bar{u}_i\left(o_j^\kappa\right) \geq \bar{u}_i\left(o_i^k\right) \\ decline, & \forall\, o_j^\kappa \in O_j^H : \bar{u}_i\left(o_j^\kappa\right) < \bar{u}_i\left(o_i^k\right) \end{cases} \tag{3.2}$$

with $i, j \in \{\mathcal{A}, \mathcal{H}\}, i \neq j$.

In other words, offers $o_j^\kappa \in O_j^{\mathrm{H}}$ are accepted by agent i if they yield a higher or equally high utility as the own current offer o_i^k, otherwise they are declined.

Bidding Strategy

The core bidding strategy of the basic negotiation model is set to be a *reaction* component to react to events in a cooperative decision making process based on an own strategy without considering the strategy of the cooperation partner, see types of human strategy determination in Section 3.1.2. However, the prospect is an additional implementation of an identification algorithm and adaptation strategy, enhancing the reaction component in the overall model towards an *identification-prediction-action* approach, see Figure 3.2. Furthermore, Section 3.1.3 limits the behavior modeling to conceding behavior only. On this basis and to ensure an agreement without the ability to define a common conflict deal with a human agent present, this work proposes the bidding strategy to be a time-based concession strategy with a continuously increasing concession [Baa16, pp. 27-28].

Hence, the concession strategy is based on a time-dependent *target utility* $\bar{u}_{\mathrm{t}}(t)$ which is decreasing over time and which the agent tries to track with the available offer utility values.

Definition 3.7 (Normalized Target Utility)
The definition of the time-dependent normalized target utility $\bar{u}_{t,i}(t) : \mathbb{R}^+ \mapsto [0,1] \subset \mathbb{R}$ *is*

$$\bar{u}_{t,i}(t) := \max_{o \in O} \left(\bar{u}_i(o)\right) \cdot \left(1 - \left(\frac{t}{\mathcal{T}}\right)^{1/\epsilon_i}\right) \tag{3.3}$$

with the concession rate $\epsilon_i \in \mathbb{R}^+$, $i \in \{\mathcal{A}, \mathcal{H}\}$ *and a negotiation deadline* $\mathcal{T} \in \mathbb{R}^+$, *assuming* $t \in [0, \mathcal{T}] \subset \mathbb{R}$.

A set of exemplary target utility trajectories for various *concession rates* ϵ with

$$\max_{o \in O} \left(\bar{u}_i(o)\right) = 1$$

is depicted in Figure 3.4.

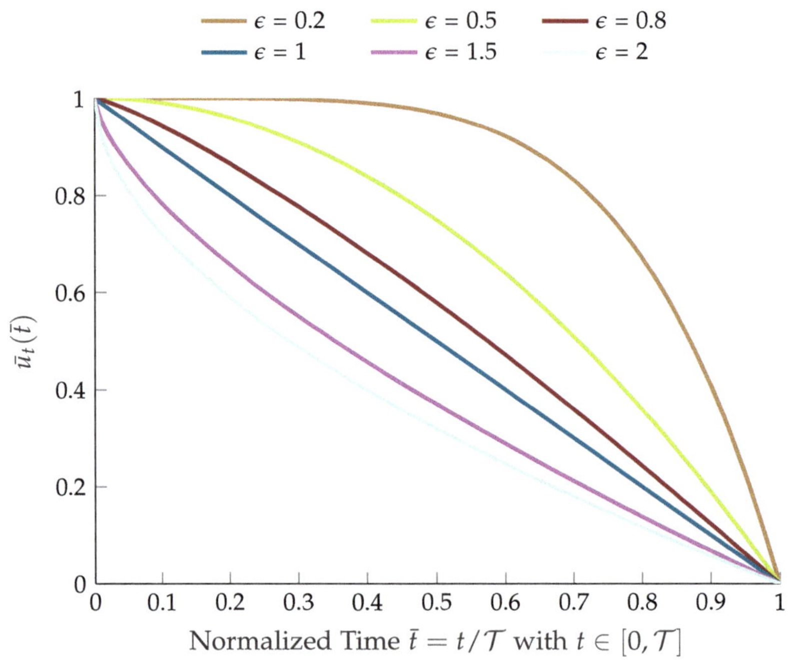

Figure 3.4: Exemplary target utility trajectories for various concession rates.

The concession of agents, i. e. the target utility tracking of agents, is defined by the following optimization problem: it determines offer o_i^t of agent i at time $t \in [0, \mathcal{T}]$ on the basis which offers' utility is closest to but greater than the current target utility.

Definition 3.8 (Concession Strategy)

The concession strategy is based on the utility Definitions 3.5 and 3.7 and determines the potentially new, best fitting offer at time $t \in [0, T]$ according to this optimization:

$$\begin{aligned} o_i^* &= \arg\min_{o \in \tilde{O}} \{\bar{u}_i(o) - \bar{u}_{t,i}(t)\} \\ s.t.\ \tilde{O} &:= \{o \in O : \bar{u}_i(o) \geq \bar{u}_{t,i}(t)\} \end{aligned} \tag{3.4}$$

If this currently best fitting offer was not yet proposed, i. e. $o_i^* \notin O_i^H$, it is offered to the other agent ($o_i^t := o_i^*$) and added to the offer history set O_i^H.

This concession definition allows for modeling various concessive negotiation behaviors, i. e. giving in linearly over time ($\epsilon = 1$), early ($\epsilon > 1$) or late ($\epsilon < 1$) with respect to to a given deadline. In literature, these concession behaviors are also called "neutral", "concessive", and "competetive",[13] respectively [VKG14]. The suitability of the defined concession strategy and target utility to model and predict human negotiation behavior was experimentally examined and confirmed in the course of research for this thesis, see detailed report in Section 4.1.

Furthermore, the above defined time-based concession strategy ensures an agreement without any conflict deal when reaching the deadline, assuming no two offers are placed at the same time instances. From a practical point of view, this assumption will be fulfilled in the intended scope of human-machine negotiation as it is nearly impossible that human and machine each propose offers at exactly the same time. However, the next section provides a necessary and sufficient criterion which also yields a theoretical agreement guarantee.

Investigations on Agreement Guarantees

The following investigations are based on a continuous-time contemplation of the basic negotiation model reasoning, which is described by the following assumption.

Assumption 3.8. *Agents perform their reasoning process of the basic negotiation model, depicted in Figure 3.3, with an infinitely small sampling time.*

For this case, the following lemma states a necessary and sufficient criterion for the uniqueness of times at which agents are proposing offers after the initial offers. Note that initial offers starting the negotiations are allowed and favored to be proposed simultaneously by both agents, see also the definition of the negotiation protocol in Definition 3.3.

13 In order to avoid confusion with the models' limitation to concessive cooperative decision making behavior, these terms are avoided in the following.

Lemma 3.1 (Criterion for the Uniqueness of Agents' Offer Timing)
After the initial offers have been placed, the times at which subsequent offers are proposed in accordance with the concession strategy of Definition 3.8 are unique for both agents if Assumption 3.8 holds and if and only if

$$\left(1 - \frac{\bar{u}_i(o_i)}{\max_{o \in O} \bar{u}_i(o)}\right)^{\epsilon_i} \neq \left(1 - \frac{\bar{u}_j(o_j)}{\max_{o \in O} \bar{u}_j(o)}\right)^{\epsilon_j} \tag{3.5}$$

holds $\forall o_i, o_j \in \{o_i, o_j \in O | o_i \neq \arg\max_{o \in O} \bar{u}_i(o), o_j \neq \arg\max_{o \in O} \bar{u}_j(o)\}$ *and for* $i, j \in \{\mathcal{A}, \mathcal{H}\}$, $i \neq j$.

Proof:
First, the uniqueness of times one agent $i \in \{\mathcal{A}, \mathcal{H}\}$ proposes new offers is shown: According to (3.1b) of the Normalized Utility Definition 3.5, one agent's utilities of all offers differ from each other. The uniqueness of times the offers are proposed follows considering the strict monotonicity of the target utility (3.3) and the unambiguity of the concession strategy in Definition 3.8 if Assumption 3.8 holds.

Second, the times t at which agents may propose new offers simultaneously after the initial offers are examined: The critical condition for an agent $i \in P$ to propose a new offer when evaluating the concession strategy continuously follows from (3.4), i. e.

$$\bar{u}_i(o_i) - \bar{u}_{t,i}(t) = 0 \quad \text{with } o_i \in \left\{ o_i \in O \,\middle|\, o_i \neq \arg\max_{o \in O} \bar{u}_i(o) \right\}. \tag{3.6}$$

Inserting the target utility definition (3.3), followed by some rearrangement yields

$$\frac{t}{T} = \left(1 - \frac{\bar{u}_i(o_i)}{\max_{o \in O} \bar{u}_i(o)}\right)^{\epsilon_i} \tag{3.7}$$

Note, that the division by $\max_{o \in O} \bar{u}_i(o)$ is legitimate due to it being non-zero which follows directly from the utility function definition and its uniqueness in Definition 3.5.

What remains is to equate the two conditions of both agents by means of the identical time t which yields (3.5). □

Note. *Criterion* (3.5) *implies that at least one agent i has to have a minimum utility of any offer which is greater than zero, i. e.*

$$\exists i \in \{\mathcal{A}, \mathcal{H}\}: \min_{o \in O} \bar{u}_i(o) > 0. \tag{3.8}$$

Upon this lemma on timing uniqueness, the following theorem states the guarantee of agents arriving at an agreement before the deadline is reached.

Theorem 3.1 (Agreement Guarantee)
Assume the criterion of Lemma 3.1 is fulfilled and Assumption 3.8 holds in the case of agents proposing new offers according to Definition 3.8. Then, it is guaranteed that an agreement is found before the deadline $\mathcal{T}$ is reached.

Proof:
What follows is a proof by contradiction. If the deadline is reached and no agreement was found, both agents would have left one final offer each with a utility of zero which has to be proposed at $t = \mathcal{T}$. This is due to the uniqueness of utilities (see Definition 3.5), the strict monotonicity of the target utility (3.3) and the unambiguity of the concession strategy in Definition 3.8 and constitutes the critical situation for the theorem.

However, both agents proposing zero-utility offers simultaneously is a violation of the criterion provided by Lemma 3.1. According to this lemma at most one agent i may reach the deadline with a zero-utility offer not proposed before the deadline. However, at that point in time the other agent j must have proposed the entire offer set and hence must have agreed on an earlier offer of agent i. □

Consequently, the fulfillment of Lemma 3.1 allows for not having a conflict deal in place.

Assumption 3.9. *The criterion introduced in Lemma 3.1 holds for the subsequent theoretical analysis of the adaptive negotiation model to guarantee an agreement of a negotiation and hence avoid the definition of a conflict deal within the model.*

Remark. *From a practical point of view, the automated agent will operate with some reasonable sampling time. In this case, the above criterion of Lemma 3.1 and Theorem 3.1 are not applicable. Therefore, the negotiation protocol implementation has to take care of assuring offer timing uniqueness. This is the motivation of restrictions in the asynchronous negotiation protocol from Definition 3.3. To guarantee that agents arrive at an agreement, the automation design may ensure that condition* (3.8) *holds for the automation, i. e. the least valuable offer's utility is greater than zero. That way, the automation will always ultimately concede, which also reflects current legislative requirements [FDM+20].*

Upon the above introduced customizations and enhancements of the basic negotiation model towards its application in human-machine negotiations in terms of asynchronous negotiation protocol, time-based concession strategy, and agreement guarantee, the following sections provide a suitable selection and application of a negotiation behavior identification approach and introduce the new explicit adaptation module of the adaptive negotiation model.

3.2.4 Identification of Negotiation Behavior

In order to influence the outcome of the negotiation, agents may use the information of the other agent's offers to identify an opponent model and apply this information within their bidding strategy. In literature, various opponent models are available, e. g. [CJ04, HT08, HL14]. Facing the challenge of little communication between automation and human within one round of negotiation (see Sections 3.1.1 and 3.1.3), a model-based identification approach capable to identify human behavior over several negotiation rounds is favored.

In the course of the research for this thesis two model-based identification approaches were considered: *nonlinear least squares* and *Bayesian learning*. In a simulative evaluation both approaches performed similarly well, although theoretically both approaches are prone to not converge or yield inconsistent results.[14] Due to the fact that Bayesian learning was designed to cope with model uncertainty, the fact that the adaptive negotiation model may not definitely represent human negotiation behavior, and the successful application of Bayesian learning in the context of human-agent negotiation (e. g. [HT08]), Bayesian learning was selected for the implementation of the adaptive negotiation model.

What follows is an application and customization of the general Bayesian learning approach to the context of the adaptive negotiation model. First, some modeling assumptions about the other agent's negotiation behavior are made.

Assumption 3.10. *For reasons of conformity and without other knowledge, it is assumed that the agents follow the same basic negotiation model and only differ in their parameters* $\boldsymbol{\theta}$ *of utility function, bidding/concession and acceptance strategy, i. e.* $\boldsymbol{\theta}$ *comprises e. g. the concession rate* ϵ *and the utility function weights* w.

With regard of a practical application, the following aspects concerning these parameters is assumed.

Assumption 3.11. *The ranges of the parameters, i. e. the ranges for each element of* $\boldsymbol{\theta}$, *are known and uniform discretization of these ranges yield suitable approximations of the actual parameters.*

14 Nonlinear least squares approaches are generally biased and may not converge due to non-convex problems. Bayesian learning may only yield consistent results for large numbers of observations according to the Bernstein-von Mises theorem [van12, pp. 138-152].

Upon these assumptions and observed offers of the other agent, Bayesian learning identifies the unknown parameters $\boldsymbol{\theta}_j$ of the other agent's utility function and bidding strategy.

Definition 3.9 (Identification Approach Based on Bayesian Learning)
Regarding Assumptions 3.10 and 3.11, a set of n_h hypotheses concerning the other agents' parameters is established by means of the discretized ranges of parameters $\boldsymbol{\theta}$:

$$H := \left[\theta_1^1, \ldots, \theta_1^{n_{h1}}\right] \times \ldots \times \left[\theta_{n_\theta}^1, \ldots, \theta_{n_\theta}^{n_{hn_\theta}}\right]$$

each resembling a specific and unique combination of parameters $\boldsymbol{h}_l \in H$ with $l \in [1, n_h]$, $n_h = n_{h1} \cdot n_{h2} \cdot \ldots \cdot n_{hn_\theta}$ and n_θ denoting the number of parameters.
For all $l \in [1, n_h]$, the initial probabilities $p^0(\boldsymbol{h}_l)$ of these hypotheses $\boldsymbol{h}_l$ are set according to a uniform distribution.
Within each iteration k of Bayesian learning at time t_k the likelihood of parameter hypotheses $p^k(\boldsymbol{h}_l)$ $\forall l$ is updated by means of the discretized Bayes' rule and the current offer o_j^κ, $\kappa < k$ of the other agent:

$$p^k\left(\boldsymbol{h}_l \mid o_j^\kappa\right) = \frac{p^k(\boldsymbol{h}_l)\, p^k\left(o_j^\kappa \mid \boldsymbol{h}_l\right)}{\sum_{x=1}^{n_h} p^k(\boldsymbol{h}_x)\, p^k\left(o_j^\kappa \mid \boldsymbol{h}_x\right)}, \tag{3.9a}$$

$$p^{k+1}(\boldsymbol{h}_l) := p^k\left(\boldsymbol{h}_l \mid o_j^\kappa\right) \tag{3.9b}$$

Considering the uniqueness of the assumed other agent's concession strategy $\mathcal{E}_j$ (see Definition 3.5), the conditional probability $p^k\left(o_j^\kappa \mid \boldsymbol{h}_l\right)$ at the current time t_k can be defined as

$$p^k\left(o_j^\kappa \mid \boldsymbol{h}_l\right) := \begin{cases} 1, & \text{if } o_j^\kappa \text{ is the result of (3.4) parameterized with } \boldsymbol{h}_l \\ 0, & \text{else.} \end{cases} \tag{3.9c}$$

The estimate of the other agent's parameters can be determined as the expected value of $\boldsymbol{\theta}$ with respect to all $\boldsymbol{h}_l$, i. e.

$$\hat{\boldsymbol{\theta}}_j^k := \sum_{l=1}^{n_h} p^k(\boldsymbol{h}_l) \cdot \boldsymbol{h}_l. \tag{3.9d}$$

To assess the uncertainty of the estimation, the standard deviation can be determined by:

$$\sigma^k := \left(\sum_{l=1}^{n_h} p^k(\boldsymbol{h}_l) \cdot \left(\boldsymbol{h}_l - \hat{\boldsymbol{\theta}}_j^k\right)^2\right)^{1/2}. \tag{3.9e}$$

The definition of the conditional probability $p^k\left(o_j^\kappa \mid \boldsymbol{h}_l\right)$ in (3.9c) is a major design issue in Bayesian learning. Commonly, the conditional probability is either set to the exact probabilities for which o_j^κ follows from $\boldsymbol{h}_l$ if these are deterministically known or they are set in accordance to a probability distribution describing a fuzzy causal relationship between o_j^κ and $\boldsymbol{h}_l$. One instantiation of the probability distribution in the latter case is the definition of a normal distribution with respect to parameters $\boldsymbol{h}_l$ given an offer o_j^κ. The expected value and variance of the normal distribution can be chosen on the basis of observed hints on the causal relationship between parameters and offers. However, due to Assumption 3.10, the conditional probability $p^k\left(o_j^\kappa \mid \boldsymbol{h}_l\right)$ is deterministically specifiable in (3.9c) given the postulated concession behavior described by (3.4) parameterized with $\boldsymbol{h}_l$. This "Dirac" definition of the conditional probability yields faster convergence than e. g. a definition relying on a normal distribution. However, the chosen form is prone to more quickly exclude hypotheses compared to the normal distribution version.

Remark. *To avoid the persistent exclusion of hypotheses with $p^k(\boldsymbol{h}_l) = 0$, especially in scenarios with agents that change their negotiation behaviors or with inadequate assumptions on the other agent's evaluation functions $\bar{b}$ (see Definition 3.5), hypotheses' probabilities can be reinitialized before each estimation update by adding a small offset q followed by normalization.*

In general, the convergence accuracy and the speed of this Bayesian learning approach also depends on the rate of observed offers of the other agent and the rate of Bayesian updates.

In practical application, this approach usually yields sufficiently fast and accurate parameter estimates. Furthermore, due to the expectation calculation (3.9d), the parameter estimate $\hat{\boldsymbol{\theta}}_j$ will be in the range of hypotheses and not diverge which is also a crucial aspect in a practical implementation.

The actual instances of Bayesian updates, i. e. the estimation updates on $\hat{\boldsymbol{\theta}}_j$, may be performed synchronously or asynchronously with the agent's basic negotiation model reasoning of Figure 3.3.

Note. *Even at times when there is no new offer of the other agent available, productive estimation updates are possible. This is due to the time-based concession strategy (see Definition 3.7 and 3.8) in which also sticking to an offer and not proposing a new offer is valuable information for parameter estimation.*

With regard to the next section introducing the explicit, generalized adaptation approach, the estimation update rate must be higher than the adaptation rate as the adaptation relies on accurate (and at best converged) estimations.

3.2.5 Explicit, Generalized Adaptation Approach

In the following the explicit adaptation component of the adaptive negotiation model is introduced. It alters the agent's parameters of the basic negotiation model from Section 3.2.3 based on the insights given by the identified negotiation behavior of the other agent, see Section 3.2.4. This way, the adaptation module enhances the reaction-based basic negotiation model towards the more advanced identification-prediction-action approach, see Section 3.1.2. Generally, behavior-sensitive approaches are favorable compared to purely time-dependent strategies as they account for individual negotiation behaviors of other agents and perform better in experiments [VKG14].

Usually, the identified negotiation behavior information is directly included in the bidding strategy [HL14], e. g. to choose an offer that suits the other agent best in case one is indifferent towards multiple potential offers [FIZ$^+$16, p. 137]. Other approaches use utility predictions to adapt the target utility and thus concession behavior with the aim to maximize utility [CATW13].

However, in this model, a more generalized adaptation principle is included which is based on an explicit evaluation of the agents' current negotiation behavior. The basis of this adaptation is the prediction of the negotiation outcome, assuming that both agents follow the basic negotiation model and that the corresponding parameters $\boldsymbol{\theta}_i$ and $\hat{\boldsymbol{\theta}}_j$ are known or estimated by agent $i \in P$.

Lemma 3.2 (Predictability of Negotiation Course and Outcome)
With a negotiation model according to Definition 3.4 and Assumptions 3.9 and 3.10, knowledge of parameters $\boldsymbol{\theta}_i$ and identification of $\hat{\boldsymbol{\theta}}_j$ according to Definition 3.9, agent $i \in \{\mathcal{A}, \mathcal{H}\}$ can predict the course and outcome of the basic negotiation model, i. e. the offer sequence, the agreed final offer and corresponding utilities.

Proof:
This statement follows trivially, considering the deterministic nature of the basic negotiation model functions (see Definition 3.4) with a unique offer timing and a guaranteed agreement (see Assumption 3.9), knowledge of all structures of these functions (see Assumption 3.10) and their (identified) parameters ($\boldsymbol{\theta}_i$, $\hat{\boldsymbol{\theta}}_j$). □

Hence, agent i is able to profit from this information by determining optimal negotiation parameters $\boldsymbol{\theta}_i^*$ with respect to an individual objective function $\mathcal{J}_i$.

Definition 3.10 (Explicit, Generalized Adaptation Approach)
The generalized adaptation approach is based on the negotiation's predictability of Lemma 3.2: considering the potential utility outcome $\bar{u}_{f,i}(\boldsymbol{\theta}_i, \hat{\boldsymbol{\theta}}_j)$ *and required effort* $\gamma_i(\boldsymbol{\theta}_i, \hat{\boldsymbol{\theta}}_j)$ *for persuading the other agent, the optimal parameters of agent* $i \in \{\mathcal{A}, \mathcal{H}\}$ *for the basic negotiation model are:*

$$\boldsymbol{\theta}_i^* = \arg\min_{\boldsymbol{\theta}_\kappa} \mathcal{J}_i\left(\bar{u}_{f,i}(\boldsymbol{\theta}_\kappa, \hat{\boldsymbol{\theta}}_j), \gamma_i(\boldsymbol{\theta}_\kappa, \hat{\boldsymbol{\theta}}_j)\right) \tag{3.10}$$

For the intended application in the context of human-machine negotiation, the effort of persuading the other agent in relation to the expected reduced loss of utility is examined. In this work, it is proposed to measure the effort of persuading by means of the time t_f from the start of a negotiation to achieving an agreement. Furthermore, only the bidding/concession strategy parameter, i. e. $\epsilon_i \in E$, is adapted, not the weights of the utility function. Hence, the negotiation *behavior* is influenced, not the *values* of agent i.

Definition 3.11 (Optimal Concession Determination)
For the scope of human-machine negotiation, the optimization problem (3.10) *of Definition 3.10 for the optimal parameters of agent* $i \in \{\mathcal{A}, \mathcal{H}\}$ *for the basic negotiation model is refined to determine the maximum optimal concession rate* ϵ_i^**:*

$$\epsilon_i^* = \max\left\{\arg\max_{\epsilon \in E} \bar{u}_{f,i}(\boldsymbol{\theta}_i, \hat{\boldsymbol{\theta}}_j) \cdot \beta^{t_f(\boldsymbol{\theta}_i, \hat{\boldsymbol{\theta}}_j)}\right\} \tag{3.11}$$
$$s.t. \quad \epsilon \rightarrow \epsilon_i \in \boldsymbol{\theta}_i.$$

$\beta \in \left]0, 1\right]$ *is an adaptation design parameter and* t_f *represents the expected time from the beginning of the negotiation to its expected end, which is depending on the parameterization of the agents, i. e.* $\boldsymbol{\theta}_i$ *and* $\hat{\boldsymbol{\theta}}_j$*.* $\bar{u}_{f,i}(\boldsymbol{\theta}_i, \hat{\boldsymbol{\theta}}_j)$ *is the corresponding loss of utility at time* t_f *for agent* i*.*

Remark. *The maximum operator in* (3.11) *is in place to achieve a unique optimal concession rate* ϵ_i^**. The choice of the maximum operator instead of the minimum operator is motivated by the association of reduced effort with more concessive behavior, i. e. higher concession rates, resulting in agents that are just relentless enough.*

Upon this prediction of the optimal concession parameter, agent i is able to adapt the current concession rate ϵ_i^k towards ϵ_i^*, taking into account the identification speed and quality in terms of the standard deviation σ^k of identification results and the risk disposition r_i (see [MHM11]) of agent i.

Definition 3.12 (Adaptation Approach for Human-Machine Negotiation)
The adaptation of the concession rate ϵ_i^k of agent $i \in \{\mathcal{A}, \mathcal{H}\}$ is based on the optimal concession rate determination of Definition 3.11, the standard deviation σ^k of the identification result and the risk disposition r_i of agent i:

$$\epsilon_i^{k+1} := \epsilon_i^k + \alpha(\sigma^k, r_i) \cdot \left(\epsilon_i^* - \epsilon_i^k\right) \tag{3.12a}$$

The risk disposition factor $r_i \in \left]0,1\right]$ is a design parameter that influences the adaptation behavior of the agent. The higher the factor the more prepared the agent is to take risks. The proposed function $\alpha\left(\sigma^k, r_i\right) \in [0,1] \subset \mathbb{R}$ evaluates the standard deviation of the current parameter estimation and balances it with the risk factor:

$$\alpha(\sigma^k, r_i) := \frac{1}{n_\theta} \sum_{l=1}^{n_\theta} \max\left(1 - \frac{\sigma_l^k}{r_i}, 0\right) \tag{3.12b}$$

n_θ is the number of estimated parameters $\hat{\boldsymbol{\theta}}_j$ (and corresponding standard deviations).

In summary, the higher the risk disposition of an agent, the faster the behavior, i. e. concession parameter, will converge to the optimal one regarding the adaptation objective, also accepting higher standard deviations of the estimated parameters.

Remark. *Since the potential adaptation of the other agent j is not explicitly considered in the introduced identification and adaptation approach of agent i, a rather high adaptation rate of agent i potentially results in an increased uncertainty in the identification and adaptation processes of both agents. Therefore, the adaptation rate from ϵ_i towards ϵ_i^* has to be sufficiently small in a practical application such that the trajectory of the concession rate ϵ_i can be considered quasi-stationary from the perspective of agent j. This has to be taken into account by both agents due to the symmetry of the discussed setup.*

Hence, the adaptation process of agent i from ϵ_i^k towards ϵ_i^* will not be at the same rate as the offer exchange. Instead, it could take place at the end of a negotiation round. That way one can think of the reactive behavior according to the basic negotiation behavior within a negotiation round as the *tactics* of negotiation and the negotiation prediction and adaptation as part of the *strategy* of negotiation, see Figure 3.2.

Note. *To ensure uniqueness of offer timing and to guarantee an agreement, the adaptation of Definition 3.12 has to obey the criterion of Lemma 3.1.*

To conclude, the adaptation module allows to model an overall negotiation behavior that factors in the other agent's behavior (see possible strategy determination in Section 3.1.2), e. g. giving in immediately if the negotiation prediction indicates a strong resistance towards the own preference or insisting on one's preference if the corresponding outcome is worth the effort. Furthermore and in contrast to existing adaptation approaches which are strongly entangled with the bidding strategy of the basic negotiation model, this approach offers increased modeling flexibility due to the fact that the adaptation strategy can be modified without changing the basic negotiation model.

In the Appendix B, an application example of the adaptive negotiation model in the context of highly automated driving is presented: It provides simulative evidence of the adaptive negotiation model's ability to cope with the challenges of cooperative decision making involving humans in terms of identifying negotiation behavior and adapting to it. Furthermore, the example also highlights the model's characteristic that offers can contain more information besides the chosen decision option. This leads to more communication and hence an increased information exchange within one round of negotiation which facilitates the identification of negotiation behavior.

In the course of negotiation theory research for this thesis, also time-variant utility functions have been investigated. An application example for negotiating driving maneuver in an evasion scenario was published [RSFH19]. However, negotiation models allowing for time-variant utility functions are not restricted to concessive behavior and complicate agreement guarantees, identification and adaptation strategies. This resulted in the limitations introduced in Section 3.1.3 for the considered cooperative decision making models in this thesis.

After the introduction of the adaptive negotiation model, the game theoretic model for emancipated human-machine cooperative decision making, the n-stage war of attrition, is presented in the next section.

3.3 The n-Stage War of Attrition

The *n-stage war of attrition* was developed in the course of two master theses [Ste18, Tan20] which led to two publications [RSFH20, RTIH20]. It advances the conventional *war of attrition* by means of a *generalized disagreement cost function* and a *stage concept*. These enhancements allow for modeling human-machine cooperative decision making soft deadlines and with multiple decision options.

To this end, the next two sections provide explanations of the required game-theoretic terminology and a review of relevant existing game-theoretic models. Subsequent sections present the n-stage war of attrition by means of introducing the stage concept and the generalized disagreement cost function.

3.3.1 Introduction and Terminology

Game theory has its origin in the mathematical modeling of strategic decision making scenarios with two or more rational entities. It is therefore qualified to be considered for the modeling of human-machine cooperative decision making as discussed in Section 3.1. If not stated differently, the following explanations are based on the work of Fudenberg and Tirole [FT91].

Game theory typically considers independent rational entities called *players*, e. g. humans, animals, societies, companies, etc., whose interactions are observed and described by means of a *game model*[15]. In contrast to automated, programmable agents in negotiation theory, players in game theory are no objects of design. Furthermore, game theory differs from conventional *decision theory*[16] by considering the influences of decision making within the decision making process, i. e. the decisions of one player depend on the decisions of all other players in the game and vice versa.

Models of game theory can be divided into two major classes, *cooperative* and *non-cooperative* games: in cooperative games, players can commit to contracts among themselves whereas in non-cooperative games all players act egoistically but still consider the decision making behavior of other players. In the context of this thesis, contracts in human-machine cooperation are not considered (see Definition 3.2) and therefore this work considers non-cooperative games only.

In a general game setup, players face decision options which usually also resemble the *actions*[17] of players. Each player values these decision options/actions individually[18] and receives a corresponding *payoff* which depends on the options chosen by himself and the other players and the *dynamicity* of the game. This dynamicity comprises two aspects: Games are either *dynamic* or *static*, depending on whether or not time or sequences of actions are considered. Furthermore, games can be *one-shot games*, if they are only played once, *repeated games*, i. e. there are several rounds of the same game, or *multi-stage games* which are sequentially interconnected non-identical games. In consequence, the time at which players receive payoffs, which may be continuously over time, at the end of a game, or dependent on the number of rounds (of the game) played, will ultimately influence the players' actions. In a realization of a game, each player's actions are determined by a *strategy* the player chooses. The strategy defines which actions will be performed depending the status of the game, especially with respect to the other players' actions or strategies. Furthermore, all strategies of one player form the player's *strategy set* and a combination of strategies with exactly one strategy for each player constitutes a *strategy profile*.

15 In the following, *game model* will be sometimes abbreviated by *game*.

16 Decision theory comprises models and approaches for rational decision making of individuals, especially in uncertain environments [Mye91, p. 5].

17 In comparison to negotiation theory, there is usually no difference between actions and offers, i. e. game theory does not provide an additional communication layer.

18 The utility of decision options in game theory is generally considered to be time-invariant in contrast to the utility of decision options in negotiation theory, which may in general be time-dependent.

Upon this general setup, the other major purpose of game theory apart from modeling strategic decision making scenarios is to provide and analyze *solution concepts*. By means of these solution concepts players are able to establish *solution strategies* to *play/solve* the game. A *solution* of a game denotes the strategy profile resulting from the corresponding solution concept. Furthermore, the establishment of strategies is not only influenced by the players' payoffs and game rules but also by the *determinism* of the game setting, the *state of knowledge* of players and their *type* or *level of rationality*.

In a deterministic game, players act and receive payoffs deterministically, i. e. there is no influence of probability on players' actions or payoffs. The human-machine cooperation setup considered in this work is assumed to be deterministic as the form of interaction and reasoning may not change, see Section 3.1.3. As a consequence, the presented approaches in this work strive to find *deterministic strategies*.

The state of knowledge describes how much players know about the game, i. e. rules and *history of actions*, as well as about other players, i. e. their payoffs, strategies and rationality. Due to the fact, that the considered use-case involves unidentified humans, players will have no or only little knowledge about the other player and therefore face an *incomplete information game*. The typical form of incomplete information is that the payoffs are private information of each player. This private information is denoted as the *type* of a player. In a realization of a game, *nature* randomly assigns players' types according to a probability distribution. This probability distribution is common knowledge upon which players form a *belief* about the other players' types. The belief may also be influenced by the history of action within a game. Ultimately, players' strategies in incomplete information games depend on the belief on the other players' types and on the potential update of this belief throughout the game.

Rationality describes the *depth of strategic thinking* in pursuing the maximization of the own payoff as elaborated on in Section 3.1.1. Generally, humans are considered to exhibit a bounded rationality, i. e. they maximize their payoff based on a finite cognitive level. Taking into account this bounded rationality of humans may be beneficial in modeling human-machine cooperation and automation design.

After determining strategies for these deterministic, incomplete information games with players of bounded rationality, the resulting strategy profile and hence the corresponding solution concept can be analyzed. If all players choose the same strategy, the resulting strategy profile is called *symmetric*. Of high importance is the persistence and stability of strategy profiles: an *equilibrium* is a solution concept in which a strategy profile is stable with respect to the game's definition (including the players' definition), i. e. no player changes the strategy despite they are generally permitted to do so. Important equilibria in the context of this thesis are defined in Appendix C.1. One famous example is the *Nash equilibrium* in which strategies of the corresponding strategy profile are *best responses* to each other with respect to the individual payoff and the other strategies in the strategy profile, see Definition C.1. Hence, no player has an incentive to change the chosen strategy.

To conclude, game theory may be a suitable framework to mathematically model the relation and strategic interaction between a human and a machine in a cooperative decision making scenario. Moreover, it provides a large toolbox for determining and analyzing strategies to take part in the process of cooperative decision making.

3.3.2 Discussion of Relevant Existing Games

In literature, various game models are available which can model certain aspects of the meta-model of human-machine cooperative decision making introduced in Section 3.1.

For example, *differential games* are often applied to model human-human or human-machine interaction on a trajectory basis since they consider the dynamics of the interaction system. Exemplarily, Na and Cole [NC15] and Flad et al. [FRDH14] base their design of driving assistance systems on differential games. In these cases, the vehicle is the interaction system and the assistance system cooperates with the driver in tracking a given reference driving trajectory of the vehicle. However, in the context of cooperative decision making, the decision options would be various differing reference trajectories. Due to the fact that solution methods for differential games assume similar, i. e. conflict-free reference trajectories, they are not suitable to resolve conflict situations, i. e. agreeing on conflict-free reference trajectories. Hence, differential games are not applicable to model and support human-machine cooperation on decision level with discrete decision options as required by the introduced corresponding meta-model, see Definition 3.2.

The *Rubinstein bargaining game* considers two players that have to split a prize by alternatingly placing offers on how to divide the prize. Due to individual discounting factors of players that reduce the players' subjective values of the prize over time, a concept of impatience is integrated into this game model. Solutions and equilibria exist both for the case of complete information [Rub82] as well as for the case of incomplete information [AG00]. Although this model is dynamic, the continuous set of decision options and the necessity for an alternating form of interaction make it unsuitable for an application in human-machine cooperative decision making, see Definition 3.2.

In the field of coordination games, Zlotkin and Rosenschein [ZR89] described a problem considering the workload distribution among postmen. They propose the extended Zeuthen strategy [Zeu19] to achieve a Nash equilibrium by iteratively and simultaneously exchanging offers in a complete information setting. For the case of incomplete information, they analyze the possibility of exchanging the relevant information before the start of the game. However, in the desired scope of cooperative decision making, the mutual exchange of all relevant information before the start of an cooperative decision making process can in general not be realized, see Assumption 3.2 and Definition 3.2.

The revision game of Calcagno et al. [CKLS14] models a deadline until which the players have to agree on an discrete action. Before reaching the deadline, players can revise their choice of action at times determined by a Poisson process. Caruana and Einav [CE08] introduce a similar model with switching costs if players change their choices of actions. Solutions to these games are provided for a complete information setting with two actions available. However, in this complete information setting, the solution strategies lead to an instantaneous agreement, i. e. there is no extended process of decision making. Therefore, both aforementioned models are unsuitable to model cooperative decision making processes in incomplete information scenarios as required in the meta-model of human-machine cooperative decision making, see Definition 3.2.

The war of attrition was proposed by Maynard Smith [May74] to model animal behavior in conflict situations with an incomplete information setting. Since then, the war of attrition has been advanced by various researchers, most of them focusing on evolution within markets or human and animal societies (e. g. oligopoly theory [FT86, BK99], establishment of technical standards [DM94], auction theory [KM97, AM06, HS11], bargaining theory [AG00], animal conflicts in evolutionary biology [BC78, BCM78, CRN12]). The original war of attrition considers two decision options and two players who pursue the goal to outlast the other player in order to win a price while facing linearly increasing costs over time if no agreement is reached. The valuation of the price is preexisting and private information of the players. The provided solution strategy for determining thresholds for giving in leads to a unique Bayesian Nash equilibrium. As a consequence, the war of attrition modeling approach is in general promising as it combines the modeling of the decision making process with the incomplete information setting. However, the majority of existing war of attrition models do only consider linear cost functions and two decision options [FT86, BK99] or players who choose valuation bids for multiple decision options in a signaling/auction game setting [DM94, KM97, AM06, HS11]. The latter manifestation of the war of attrition model is not suitable to model human-machine cooperation on decision level as the valuations of decision options should be in a predefined relation to the decision options, see Definition 3.2. Otherwise it would be generally unclear how an automation should establish its valuations of decision options. However, the first manifestations of the conventional war of attrition model offers some promising starting points for taking into account the discussed requirements of human-machine cooperative decision making, see Definition 3.2.

Therefore, an enhancement of the conventional war of attrition model towards the generalization of the original time-linear cost function to a strictly increasing time-dependent cost function and the consideration of more than two decision options is introduced in the following sections. As a first step in modeling human-machine cooperative decision making by means of the war of attrition concept, the game model applied in the context of this thesis is defined. It is customized to suit the requirements of human-machine cooperation on decision level summarized in the corresponding meta-model Definition 3.2 and hence differs from the conventional war of attrition definition by allowing for multiple decision options and an arbitrary but strictly increasing disagreement cost function, resembling a shapeable soft deadline. In a second step, solution strategies for the applied game model and corresponding equilibria are presented.

3.3.3 The Applied Game Model of the War of Attrition

The applied game model of the war of attrition generalizes the conventional model (see [May74, FT91]) towards the requirements of human-machine cooperation on decision level (see Assumptions 3.1, 3.4, and 3.5) by allowing for multiple decision options and (soft) decision making deadlines in form of increasing disagreement cost functions. The game model is defined as follows.

Definition 3.13 (The Applied War of Attrition Game Model)
The applied war of attrition game model is described by the tuple $\mathcal{G}(P, D, c, U, \Pi, \mathcal{F})$:

- *A set of two rational players P, $|P| = 2$.*
- *A discrete set of decision options D that is identical for both players and to the set of offers $D \equiv O$.*
- *A time-dependent, cumulative cost function $c(t) : \mathbb{R}^+ \mapsto \mathbb{R}^+, c(0) = 0$, which is common knowledge.*
- *A set of utilities $U_i \subset U \subset \mathbb{R}$ as the basis for each player $i \in P$ to value every decision option $d_i \in D$ with an individual utility $u_{d_i} \in U_i$, i.e. each utility $u_{d_i} \in U_i$ is unique and $|U_i| = |D|$. The utility information U_i is private knowledge and resembles the type of player i.*
- *A set of payoff functions Π for both players mapping any pair of decision options to a utility of U reduced by the disagreement costs at time t:*

$$\pi_i \in \Pi_i \subset \Pi : D \times D \times \mathbb{R}^+ \mapsto \mathbb{R}$$
$$\pi_i(d_i, d_j, t) := \begin{cases} u_{d_i} - c(t), & d_i = d_j \\ -\infty, & d_i \neq d_j \end{cases} \tag{3.13}$$

  with $u_{d_i} \in U_i$, $d_i, d_j \in D$ and $i, j \in P$, $i \neq j$.
  The objective of both players is to maximize their individual payoff.

- *A set of probability density functions $\mathcal{F}$ with $f_{\delta_i}(\delta_i) : \Delta_i \mapsto \mathbb{R}^+$, $f_{\delta_i} \in \mathcal{F}\ \forall i \in P$, which are non-zero except for $\delta_i = 0$, i.e. $f_{\delta_i}(0) = 0$, and $\delta_i \to \infty$, i.e. $\lim_{\delta \to \infty} f_{\delta_i}(\delta) = 0$. The utility difference $\delta_i \in \Delta_i \subset \mathbb{R}^+$ is defined as the difference between neighboring elements of the ordered set $\vec{U}_i$ which is the set U_i of player $i \in P$ with elements in descending order and with $|\Delta_i| = |\vec{U}_i| - 1$. The probability density function f_{δ_i} as well as the corresponding cumulative distribution function $F_{\delta_i}(\delta_i) : \mathbb{R}^+ \mapsto [0, 1]$, $i \in P$, are common knowledge.*

The rules of the game are: The game starts at time $t = 0$ with an initial decision option offer of both players. If the initial offers are equal, the game ends immediately and players receive their payoffs. Otherwise, the game continues and both players $i \in P$ are able to place new offers of decision options, both establishing a history of decision option offers D_i^H and D_j^H. An agreement is reached as soon as $D_i^H \cap D_j^H = d_f \neq \emptyset$, $i, j \in P$, $i \neq j$, and hence the game ends at time t_f at which the offer $d_f \in D$ is placed by the player who places this offer last.

Note, that in this setup the repetition of offering one decision option which was already proposed by the same player has no influence on the payoff. Therefore, it is assumed without loss of generality that each player proposes a specific decision option offer at most once.

Considering this game model, the question to be answered next is at which times players will concede and offer subjectively less valuable decision options. These (potentially multiple) time thresholds τ_i resemble the strategy ψ_i of player i and a corresponding strategy profile represents a solution of the game.

The following sections introduce two solution approaches to different forms of the above introduced game model. At first, a game setup according to Definition 3.13 with two decision options ($|D| = 2$) is considered to focus on strategy determination with respect to the generalized disagreement cost function $c(t)$.

3.3.4 Solution Strategy for Generalized Costs

The following solution strategy for the war of attrition with two decision options ($|D| = 2$) and a generalized cost function advances the work on the conventional war of attrition, see [FT91, pp. 216-219] and [BK99]. It was developed in the course of the supervised thesis of Steinkamp [Ste18] and published afterwards [RSFH20].

The purpose of the generalized cost function is to create a cooperative decision making pressure (see Definition 3.2) which can be motivated by increasingly concessive human behavior when approaching a critical point or deadline for decision making [SGC98].

Therefore, the disagreement cost is modeled as an external, systematic influence on cooperative decision making of all players. Hence, the disagreement cost function is set to be identical and common knowledge for all players. Furthermore, the cost function is modeled as time dependent and cumulative, i. e. strictly increasing over time. Therefore and for mathematical reasoning purposes, the cost function $c(t)$ in this thesis has to fulfill the following assumption:

Assumption 3.12. *The disagreement cost function $c(t) \in C^1 : \mathbb{R}^+ \mapsto \mathbb{R}^+$ is continuously differentiable and strictly increasing, i. e. $\dot{c}$ exists and $\dot{c}(t) > 0\ \forall t \in \mathbb{R}^+$.*

As a consequence, the cost function allows for modeling a soft deadline: If the cost function becomes sufficiently steep at some point, players' utilities are not worth the effort of not conceding as both players try to maximize their payoff (3.13), and hence an agreement is reached at that point.

What follows is the introduction of a solution strategy and the corresponding equilibrium considering these kinds of cost functions.

Strategy Determination

The non-trivial part of the strategy determination is concerned with the case of a conflict situation, i. e. players initially prefer different decision options. In this situation, players try to outlast the other player. However, as all players face the same disagreement costs, players will concede and offer their second preferred decision option after some time which leads to an agreement. The times at which the players $i \in P$ concede are denoted as thresholds τ_i. These thresholds depend on player i's own utility difference δ_i and on the other player's utility difference δ_j $(i, j \in P,\ i \neq j)$, as well as on the disagreement costs $c(t)$. The function determining the thresholds dependent on these parameters is called threshold function $\tau_i(\cdot)$ and forms the basis of the players' strategies to outlast the other player. As the utility of decision options is private information, each player i has to maximize expected payoff π_i with respect to the given utility difference distribution $f_{\delta_j}(\delta)$ of the other player j in order to find the trade-off between cost and reduced loss of utility in this incomplete information setting.

Given a cost structure according to Assumption 3.12, the following assumption on the threshold function parameterized with the player's utility difference is made:

Assumption 3.13. $\tau_i(\delta) : \mathbb{R}^+ \mapsto \mathbb{R}^+$ *is strictly increasing and hence invertible. Furthermore,* $\tau_i(\delta)$ *is continuously differentiable and*

$$\tau_i(0) = 0, \tag{3.14}$$

see [FT91, pp. 216-217].

Based on the Assumptions 3.12 and 3.13, the following lemma provides the threshold function for maximizing the expected payoff.

Lemma 3.3 (Threshold Function for Generalized Costs)
Let Assumptions 3.12 and 3.13 hold. Then, the threshold of player $i \in P$ *in a war of attrition with two decision options* ($|D| = 2$) *maximizing her or his expected payoff with respect to the density distribution of utility difference* f_{δ_j} *of player* j*, a cost function* $c(t)$ *and player* i*'s utility difference* δ_i *is given by*

$$\tau_i(\delta_i) = c^{-1}\left(\int_0^{\delta_i} \tilde{\delta} \frac{f_{\delta_j}(\tilde{\delta})}{1 - F_{\delta_j}(\tilde{\delta})}\, \mathrm{d}\tilde{\delta}\right). \tag{3.15}$$

In the following, the crucial steps in deriving (3.15) are briefly presented as the approach is inspired by Fudenberg and Tirole [FT91, pp. 216-219].

Proof:
By means of Assumption 3.15, utility differences δ can be mapped to thresholds τ $\forall i \in P$. Therefore, the common knowledge utility difference distribution f_{δ_j} of the other player j can be virtually transformed into the corresponding threshold density distribution f_{τ_j}. Hence, the objective function $\mathcal{J}_i$ for maximizing the expected payoff is set up in the threshold, i. e. time, domain by means of the threshold density distribution f_{τ_j} with $f_{\tau_j}: \mathbb{R}^+ \mapsto \mathbb{R}^+$, $f_{\tau_j}(0) = 0$ and $\lim_{\tau \to \infty} f_{\tau_j}(\tau) = 0$. Furthermore, the objective function $\mathcal{J}_i$ depends on the sought threshold τ_i of player i, player i's utility difference δ_i and the cost function $c(t)$, i. e.

$$\mathcal{J}_i(\tau_i) := \underbrace{\int_0^{\tau_i} (\delta_i - c(\tau)) f_{\tau_j}(\tau)\,\mathrm{d}\tau}_{\text{expected payoff gain if player } i \text{ wins}} + \underbrace{\int_{\tau_i}^{\infty} (-c(\tau_i)) \cdot f_{\tau_j}(\tau)\,\mathrm{d}\tau}_{\text{expected payoff loss if player } i \text{ loses}} . \tag{3.16}$$

With the derivative of $\mathcal{J}_i$ by τ_i the necessary condition for the maximum is found:

$$\delta_i \cdot f_{\tau_j}(\tau_i) - \dot{c}(\tau_i)\left(1 - F_{\tau_j}(\tau_i)\right) = 0. \tag{3.17}$$

The sufficiency of condition (3.17) is the result of Lemma C.1.

According to the fundamental theorem of calculus and with Assumption 3.13, the density distribution and the cumulative distribution function of (3.17) can be transformed according to Lemma A.2 and rearranged to

$$\dot{c}(\tau_i(\delta_i))\,\frac{\mathrm{d}\tau_i}{\mathrm{d}\delta_i}(\delta_i) = \delta_i \cdot \frac{f_{\delta_j}(\delta_i)}{1 - F_{\delta_j}(\delta_i)}. \tag{3.18}$$

The transformed condition (3.18) is integrated with respect to δ_i taking into account the cost function's initial value from Definition 3.13.

$$c(\tau_i(\delta_i)) = \int_0^{\delta_i} \tilde{\delta}\,\frac{f_{\delta_j}(\tilde{\delta})}{1 - F_{\delta_j}(\tilde{\delta})}\,\mathrm{d}\tilde{\delta}. \tag{3.19}$$

Finally, with the cost function $c(t)$ being continuous, strictly increasing and therefore invertible (Assumption 3.12) and with (3.14), the threshold function (3.15) follows. □

Remark. *The threshold function* (3.15) *fulfills Assumption 3.13 of* $\tau_i(\delta)$ *being an invertible function due to the fact that the cost function is invertible (Assumption 3.12) and the integral is strictly increasing, hence also invertible. The latter results from an always positive integrand that diverges [Rin14, p. 12]. Besides this, it is easy to see that the threshold function* (3.15) *is differentiable yielding an integrable derivative.*

Having established the threshold function maximizing the player's expected payoff and assuming no differences between players in terms of rationality, the following symmetric strategy profile can be defined.

Definition 3.14 (War of Attrition Strategy Profile for Generalized Costs)
For both players $i \in P$:

1. *Start the game with offering the preferred decision option, i. e. the option with the highest utility.*
2. *If the other player does not give in before the own threshold provided by* (3.15) *is reached, concede by offering the other decision option.*

What follows is the equilibrium analysis of the above introduced strategy profile.

Equilibrium

The following theorem states the equilibrium resulting from the strategy profile of Definition 3.14.

Theorem 3.2 (Bayesian Nash Equilibrium)
The symmetric strategy profile of Definition 3.14 yields a Bayesian Nash equilibrium.

Proof:
According to Definition C.2 of the Bayesian Nash equilibrium, it has to be shown that the proposed strategy is a best response to itself considering the maximization of expected payoffs with respect to the probability for potential types of the other player. This is done by considering separately the two cases of how the game can possibly start with: If both players prefer the same option, an agreement is reached immediately without any costs. If players prefer different options, both will realize the conflict and hence the war of attrition they are in. By following the above introduced symmetric strategy of conceding only if their thresholds are reached, they individually maximize their expected payoff at all times taking into account the other player's potential types. Hence, both players find themselves in a Bayesian Nash equilibrium which consequently also applies for the overall game. □

The next section advances the introduced war of attrition with generalized costs to games with more than two decision options, i. e. $|D| > 2$.

3.3.5 Extension Towards Multiple Decision Options

In human-machine cooperative decision making, participants usually face more than two decision options. Therefore, this section provides a solution approach for the applied war of attrition game model of Definition 3.13 that is capable of handling cases with more than two decision options ($|D| > 2$) and also allows for a generalized cost function as introduced in Section 3.3.4.

The requirements of Definition 3.2 in terms of incomplete information and concessive behavior of the cooperation partners are still valid for more than two decision options. However, in the case with more than two decision options the uncertainty of players increases with respect to the unknown preference sequence of the other players, i. e. players do not only not know when the other player concedes but they are additionally unaware which decision option the other player concedes to. Hence, it is assumed that players facing disagreement costs in these conflict situations are iteratively closing in on the agreement by conceding and offering multiple other decision options. The objective of the following advanced war of attrition game model is to describe this process and the corresponding behavior of players.

One important aspect in modeling the conceding process by means of game theory is the absence of a clear interaction order, see meta-model of human-machine cooperative decision making in Definition 3.2. In other words, players are not forced to interact simultaneously or alternatingly in the course of the game. Another crucial aspect is the fact that players are able to react to the observation of the conceding behavior of the other player, i. e. when the other player proposes which other decision options. These aspects require careful consideration when determining the strategy of players for the given game setting.

As before, a function providing the thresholds τ_i, at which the player $i \in P$ concedes, forms the basis of the player's strategy of how to outlast the other player. The thresholds τ_i should still depend on the disagreement costs $c(t)$, players' own utility differences δ_i for the corresponding decision options and the utility differences δ_j of the other player ($i, j \in P$, $i \neq j$). Due to the incomplete information setting, the utility of decision options is again private information and therefore each player i has to maximize the expected payoff with respect to the given utility difference distribution $f_{\delta_j}(\delta)$ and the potentially observed conceding behavior of the other player j.

To account for these requirements, the following sections first introduce the *stage concept* for modeling the iterative closing in on an agreement. Upon this concept, the strategy that maximizes the expected payoff is derived. Furthermore, it is proven that the corresponding symmetric strategy profile yields a *perfect Bayesian equilibrium*, see Definition C.3.

Stage Concept

The introduction of the *stage concept* is motivated by the following two major aspects of the challenge to determine a solution strategy for the war of attrition with multiple decision options.

1) **The Bounded Rationality of Humans**
   Considering the game model with multiple decision options, humans face the complex task of taking into account the future course of the incomplete information game when determining their strategies. In other words, they have to anticipate which decision option the other player is proposing next and when. In line with the rationality discussion in Section 3.1.1 and supporting experiments [Nag95, CHC04, CGC06, CGIC09] showing that humans usually operate on decision level with a low level of rationality, humans may not be able to fully predict the future course of the game in all aspects and (re-)act accordingly. Instead it is assumed that humans will focus more on the current situation, history and immediate future course of the game.

2) **Analytical and Scalable Strategy Determination**
   For the practical implementation of any solution strategy, it is beneficial that its determination can be performed analytically and is scalable with respect to the number of decision options. In the course of this work, it has become obvious that considering every future course of the proposed game setup and especially future offers of decision options of the other player does not yield analytical solutions and requires numerical solutions instead. Furthermore, the input for the numerical solution methods scales poorly with respect to the number of decision options and becomes almost unmanageable when a game considers more than three decision options. [Tan20]

Therefore, it is advisable to restrain the basis of strategy determination to the history, current state and immediate future course of the game which is the objective of the stage concept. To this end, the potential iterative closing in on the agreement is split into rounds of cooperative decision making. The subsequent splitting within the game model yields a *multi-stage game*. Consequently, the rounds of cooperative decision making are called *stages*. An exemplary stage setting is depicted in Figure 3.5.

> **Definition 3.15 (Stage in the n-Stage War of Attrition)**
> *A stage $m \in \{1, \ldots, n\} \subset \mathbb{N}^{>0}$ is defined as the time span $(t_{m-1}, t_m]$ during which players do not offer new decision options. Consequently, at stage changes, i. e. at times t_m, one player offers a new decision option. The actual stage number for reaching an agreement is denoted by n. The upper boundary of stage numbers is $n \leq |D| - 1$ which is common knowledge.*

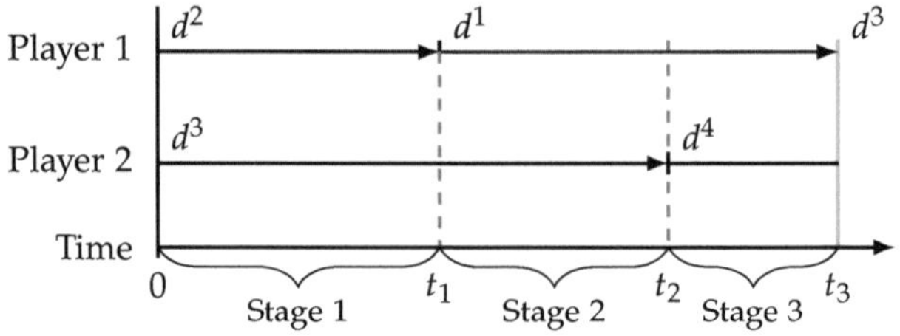

Figure 3.5: An exemplary stage setting with two players $P = \{1,2\}$, decision options $D = \{d^1, d^2, d^3, d^4\}$ and utilities $U_1 = [3,4,1,0]$, $U_2 = [0,1,4,2]$. The green vertical line on the right indicates the agreement after the third stage.

The upper boundary of stage numbers results from both players acting (at least to some degree) rational and therefore only conceding, them relying on the identical set of decision options D and each offer appearance being unique with respect to the individual player.

Note, that from the point of view of player i not every stage m that is determined by the other player offering a new decision option, i. e. giving in, provides a decision option $d_j \in D$ that is closer to the own decision option offers D_i^{H} in terms of utility compared to the other player's offer history D_j^{H}. In the example illustrated in Figure 3.5, this is the case at t_1 when player 1 newly offers d^1. From the perspective of player 2, this new offer provides an decision option with even less utility than the previously offered decision option d^2 (0 vs. 1).

Due to the incomplete information setting, players are not aware of the other player's preference sequence of the decision options in addition to corresponding unknown utility differences. Therefore, players cannot foresee the conceding sequence until an agreement is reached. Hence, the actual number of stages needed to reach an agreement in a realization of the game, i. e. n, is a priori unknown to players. This actual number of stages motivates the name *n-stage war of attrition* for this model of cooperative decision making.

To account for the unknown number of stages until an agreement is reached and considering the bounded rationality of humans discussed above, the stage concept also comprises the following assumption for strategy determination that restrains players not to consider every possible future course of the game.

Assumption 3.14. *Each player treats the current stage of the game as if the game terminates at the end of the current stage.*

Assumption 3.14 furthermore enables the following analytical determination of the solution strategy for the war of attrition with multiple decision options.

Strategy Determination

The threshold determination of each player in the war of attrition with multiple decision options is based on the maximization of the expected payoff as in the case of two decision options, see Section 3.3.4. However, the determination of the next threshold is performed at the beginning of every stage. In so doing, players follow Assumption 3.14 and only consider the current situation, game's history and immediate future course, i. e. the upcoming stage, instead of taking into account all unknown potential future courses of the game.

As one consequence, it is essential that players update their belief about the other player, i. e. the density function of utility differences, with the information they encounter in previous stages:

- The player that has not given in at the previous stage $m-1$, the *winner* of that stage, may be able to determine the utility difference which led the other player j to concede in stage $m-1$. However, due to the fact that the utility difference of player j in stage m is independent of the one in the previous stage, this information does not add value for player i in stage m.
- The player that has given in at the previous stage $m-1$, the *loser* of that stage, receives the information that the current utility difference of the other player j is greater than the own utility difference in stage $m-1$. On this basis, player i is able to adapt the utility density function $f_{\delta_j^m}$ for stage m.

Remark. *At the first stage $m = 1$ both players see themselves as winners of the previous virtual stage $m = 0$. The same applies for situations in which both players determine the same threshold and give in simultaneously.*

Furthermore, considering Assumption 3.14, the proposed solution strategy for the *n-stage war of attrition* is based on individual threshold functions $\tau_i^m(\delta) : \mathbb{R}^+ \mapsto \mathbb{R}^+$, $i \in P$, for every stage $m \in \{1, \ldots, n\}$. For these threshold functions the following is assumed similarly to Assumption 3.13.

Assumption 3.15. *$\forall i \in P$, $m \in \{1, \ldots, n\}$ the threshold function $\tau_i^m(\delta) : \mathbb{R}^+ \mapsto \mathbb{R}^+$ is strictly increasing and hence invertible. Hence, its inverse $\delta = \phi_i^m(\tau)$ exists. Furthermore, $\tau_i^m(\delta)$ is continuously differentiable and*

$$\tau_i^m(0) := 0. \tag{3.20}$$

Before presenting the analytical threshold functions maximizing the players' expected payoffs for the n-stage war of attrition solution strategy, the following notations and a lemma on players' expected payoff in each stage m are introduced.

First, let $\tau^{1:m}$ denote the time at which stage $m \in \{1, \dots, n\}$ starts:

$$\tau^{1:m} := \sum_{\kappa=1}^{m-1} \min\left(\tau_i^{\kappa},\, \tau_j^{\kappa}\right) \text{ with } \tau^{1:1} := 0. \tag{3.21}$$

Similarly, $\delta_i^{l:m}$ describes the sum of utility differences δ_i from stage l to $(m-1)$:

$$\delta_i^{l:m} := \sum_{\kappa=l}^{m-1} \delta_i^{\kappa}. \tag{3.22}$$

Furthermore, l references the last stage before the current stage m at which players' roles (winner/loser) changed, i. e. $\tau_i^l \gtrless \tau_j^l$ and $\tau_i^{(l-1)} \lessgtr \tau_j^{(l-1)}$ holds. Initially, l is set to $l = 1$.

By means of these definitions, the expected payoff of player i at stage m depending on the sought threshold τ_i^m can be stated. For simplicity, the expected payoff is firstly formulated in the threshold, i. e. time, domain by means of a threshold density distribution $f_{\tau_j^m}$ of the other player j in stage m with $f_{\tau_j^m} : \mathbb{R}^+ \mapsto \mathbb{R}^+$, $f_{\tau_j^m}(0) = 0$ and $\lim_{\tau\to\infty} f_{\tau_j^m}(\tau) = 0$.

Lemma 3.4 (Expected payoff at Stage m)
Let Assumptions 3.14 and 3.15 hold for any player $i \in P$. The expected payoff $\mathcal{J}_i^m$ for stage m is:

$$\begin{aligned}\mathcal{J}_i^m(\tau_i^m, \delta^m, c^m) := &\int_0^{\tau_i^m} \left(\delta^m - c\left(\tau^{1:m} + \tau\right) + c^m\right) \cdot f_{\tau_j^m}(\tau)\,\mathrm{d}\tau \\ &+ \int_{\tau_i^m}^{\infty} \left(-c\left(\tau^{1:m} + \tau_i^m\right) + c^m\right) \cdot f_{\tau_j^m}(\tau)\,\mathrm{d}\tau\end{aligned} \tag{3.23a}$$

with utility differences

$$\delta^m := \begin{cases} \delta_i^l & \text{if player } i \text{ has won the previous stages since stage } l < m, \\ \delta_i^m & \text{if player } i \text{ has lost the previous stage,} \end{cases} \tag{3.23b}$$

and cost functions offsets

$$c^m := \begin{cases} c\left(\tau^{1:l}\right) & \text{if player } i \text{ has won the previous stages since stage } l < m, \\ c\left(\tau^{1:m}\right) & \text{if player } i \text{ has lost the previous stage.} \end{cases} \tag{3.23c}$$

Proof:
Due to Assumption 3.15, a mapping of utility difference δ to threshold τ_i^m exists for all stages m and $\forall i \in P$. This enables the virtual transformation of the common knowledge utility difference distribution f_{δ_j} into the corresponding threshold density distribution $f_{\tau_j^m}$. Therefore, the expected payoff of player i can be formalized by means of the threshold density distribution $f_{\tau_j^m}$. Since the expected payoff depends on whether player i wins ($\tau_i^m > \tau_j^m$) or loses ($\tau_i^m < \tau_j^m$) the current stage m, the expectation integral over $f_{\tau_j^m}$ is split into these two parts, assuming that the game will end after the current stage m, see Assumption 3.14:

$$\mathcal{J}_i^m(\tau_i^m, \delta^m, c^m) := \underbrace{\int_0^{\tau_i^m} \left(\delta^m - c\left(\tau^{1:m} + \tau\right) + c^m\right) \cdot f_{\tau_j^m}(\tau)\, \mathrm{d}\tau}_{\text{expected payoff gain if player } i \text{ wins}}$$
$$+ \underbrace{\int_{\tau_i^m}^{\infty} \left(-c\left(\tau^{1:m} + \tau_i^m\right) + c^m\right) \cdot f_{\tau_j^m}(\tau)\, \mathrm{d}\tau}_{\text{expected payoff loss if player } i \text{ loses}}.$$

The first integral resembles the expected payoff gain if player i wins stage m and hence the game, see Assumption 3.14. In this case, she or he gains compared to the next smaller utility of $\vec{U}_i$ the utility difference δ^m minus the disagreement costs $c\left(\tau^{1:m} + t\right)$. The second integral yields the expected payoff loss in case player i loses the current stage, i. e. compared to the next smaller utility of $\vec{U}_i$ she or he faces the disagreement costs $c\left(\tau^{1:m} + \tau_i^m\right)$ at the end of stage m.

δ^m describes the utility difference of the current stage m. This utility difference depends on whether player i has won or lost previous stages since stage l, see (3.23b). If she or he has won, the utility difference δ_i^l of stage l is still relevant ($\delta^m = \delta_i^l$). If she or he lost the previous stage, she or he considers the new utility difference δ_i^m of stage m ($\delta^m = \delta_i^m$). In order to properly compare utility win and disagreement costs, c^m is required for a cost offset correction of the current stage m (see Figure 3.6) depending on whether player i has lost or won previous stages, see (3.23c).

□

Note. *Although the expected payoff $\mathcal{J}_i^m$ depends on τ_i^m, δ^m and c^m, the utility difference δ^m and the cost function offset c^m are determined by the specific stage setting. From the perspective of player i, only the threshold is variable, i. e. $\mathcal{J}_i^m\left(\tau_i^m\right)$.*

Having established the expected payoff $\mathcal{J}_i^m$ of player $i \in P$ for each stage $m \in \{1, \ldots, n\}$ in Lemma 3.4 the following theorem provides the threshold function that maximizes $\mathcal{J}_i^m$.

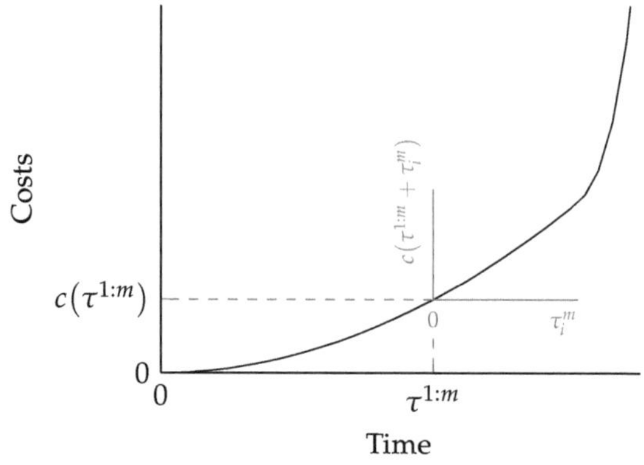

Figure 3.6: Offset correction in exemplary cost function at $\tau^{1:m}$ for player i.

Theorem 3.3 (n-Stage War of Attrition Threshold Function)
Let Assumptions 3.12, 3.14 and 3.15 hold. The expected payoff $\mathcal{J}_i^m(\tau_i^m)$ of player i (see Lemma 3.4) is maximized for all stages $m \leq n$ by the following threshold τ_i^m depending on whether player i has won or lost the previous stage $(m-1)$:

$$\tau_i^m\left(\delta_i^l\right) = c^{-1}\left(\int_0^{\delta_i^l} \tilde{\delta} \cdot \frac{f_{\delta_j}(\tilde{\delta})}{1-F_{\delta_j}(\tilde{\delta})}\,\mathrm{d}\tilde{\delta} + c\left(\tau^{1:l}\right)\right) - \tau^{1:m} \tag{3.24a}$$

if player i has won since stage l including stage $(m-1)$, otherwise

$$\tau_i^m(\delta_i^m) = c^{-1}\left(\int_0^{\delta_i^m} \tilde{\delta} \cdot \frac{f_{\delta_j}\left(\delta_i^{l:m}+\tilde{\delta}\right)}{1-F_{\delta_j}(\delta_i^{l:m}+\tilde{\delta})}\,\mathrm{d}\tilde{\delta} + c\left(\tau^{1:m}\right)\right) - \tau^{1:m} \tag{3.24b}$$

if player i has lost stage $(m-1)$.

Note. *It can be easily shown that the conventional war of attrition with two decision options has only one stage ($n=1$) with $m=l=1$ and both players considering* (3.24a).

Proof:
The two cases of the strategy definition given in Theorem 3.3 are discussed separately. The case in which player i has won in the previous stage or the game has just started ($m=1$) is considered first:

Recall the expected payoff function (3.23a) of Lemma 3.4 and the relevant case (player i has won stage $m-1$) of (3.23b) and (3.23c). Considering the Definition A.1 of integrals with infinite integration limits and following the rule for differentiation of limits of integrals (see Lemma A.1), the partial time derivative of the expected payoff function $\mathcal{J}_i^m(\tau_i^m, \delta^m, c^m)\big|_{\delta^m=\delta_i^m, c^m=c(\tau^{1:l})}$ with respect to τ_i^m can be obtained which yields the necessary condition for a maximum payoff:

$$\delta_i^l \cdot f_{\tau_j^m}(\tau_i^m) - \frac{\partial c(\tau^{1:m} + \tau_i^m)}{\partial \tau_i^m} \cdot \left(1 - F_{\tau_j^m}(\tau_i^m)\right) = 0. \tag{3.25}$$

The proof of sufficiency of condition (3.25) is analogous to Lemma C.1.

The subsequent goal is to retrieve a threshold function $\tau_i^m\left(\delta_i^l\right)$ from condition (3.25). Therefore, (3.25) is rearranged:

$$\delta_i^l \cdot \frac{f_{\tau_j^m}(\tau_i^m)}{1 - F_{\tau_j^m}(\tau_i^m)} = \frac{\partial c(\tau^{1:m} + \tau_i^m)}{\partial \tau_i^m}. \tag{3.26}$$

This rearrangement is possible for finite threshold values due to the fact that only for $\tau_i^m \to \infty$ follows $\left(1 - F_{\tau_j^m}(\tau_i^m)\right) \to 0$ which is a direct consequence of the general definition of density functions and Assumption 3.15.

Next, $f_{\tau_j^m}$ and $F_{\tau_j^m}$ in (3.26) are transformed into $f_{\tau_j^l} \equiv f_{\tau_j}$ and $F_{\tau_j^l} \equiv F_{\tau_j}$ by the argument shift of $\tau^{l:m}$ to take into account the history of victories in previous stages, i. e. past thresholds since stage l. Taking also into account Lemma A.2 for the transformation of density functions, this results in:

$$\delta_i^l \cdot \frac{f_{\tau_j}\left(\tau^{l:m} + \tau_i^m\right)}{1 - F_{\tau_j}(\tau^{l:m} + \tau_i^m)} = \frac{\partial c(\tau^{1:m} + \tau_i^m)}{\partial \tau_i^m}. \tag{3.27}$$

At this point, the virtual transformation of the proof of Lemma 3.4 is reversed, i. e. f_{τ_j} and F_{τ_j} are re-transformed to f_{δ_j} and F_{δ_j}, respectively, by means of the following mapping:

$$\delta_i^l = \phi_i^m\left(\tau^{l:m} + \tau_i^m\right), \quad \tau^{l:m} \text{ const}, \tag{3.28}$$

which resembles the inverted threshold function $\phi_i^m(\tau)$ of Assumption 3.15 in case player i has won since stage l.

Considering again Lemma A.2 for the transformation (3.28) of the density function and its cumulative distribution function, (3.27) can be reformulated as:

$$\delta_i^l \cdot \frac{f_{\delta_j}\left(\delta_i^l\right)}{1 - F_{\delta_j}(\delta_i^l)} \cdot \frac{1}{\frac{\mathrm{d}\tau_i^m(\delta_i^l)}{\mathrm{d}\delta_i^l}} = \frac{\partial c(\tau^{1:m} + \tau_i^m)}{\partial \tau_i^m} \tag{3.29}$$

Multiplying this transformed condition (3.29) with the derivative of the inverse transformation $\frac{\mathrm{d}\tau_i^m(\delta_i^l)}{\mathrm{d}\delta_i^l}$ results in:

$$\frac{\partial c(\tau^{1:m}+\tau_i^m)}{\partial \tau_i^m} \cdot \frac{\mathrm{d}\tau_i^m\left(\delta_i^l\right)}{\mathrm{d}\delta_i^l} = \delta_i^l \cdot \frac{f_{\delta_j}\left(\delta_i^l\right)}{1-F_{\delta_j}(\delta_i^l)} \tag{3.30}$$

Equation (3.30) is then integrated with respect to δ_i^l by reversing the chain rule of differentiation and considering the initial offset of (3.20):

$$c\left(\tau_i^m\left(\delta_i^l\right)+\tau^{1:m}\right) = \int_0^{\delta_i^l} \tilde{\delta} \cdot \frac{f_{\delta_j}(\tilde{\delta})}{1-F_{\delta_j}(\tilde{\delta})}\, \mathrm{d}\tilde{\delta} + c\left(\tau^{1:l}\right) \tag{3.31}$$

Due to Assumption 3.12 $c(t)$ is continuous, strictly increasing and therefore invertible, the threshold function (3.24a) results by rearranging (3.31).

The second case of (3.24b) can be proven analogously to (3.24a). Therefore, only the relevant steps and reasonings are provided. The derivative of (3.23a) with respect to τ_i^m and with $\delta^m = \delta_i^m$, $c^m = c\left(\tau^{1:l}\right)$ yields the necessary and sufficient condition

$$\delta_i^m \cdot f_{\tau_j^m}(\tau_i^m) - \frac{\partial c(\tau^{1:m}+\tau_i^m)}{\partial \tau_i^m} \cdot \left(1-F_{\tau_j^m}(\tau_i^m)\right) = 0. \tag{3.32}$$

Then, the transformation

$$\delta_i^{l:m} + \delta_i^m = \phi_i^m(\tau_i^m) \tag{3.33}$$

is introduced to re-transform $f_{\tau_j^m}$ and $F_{\tau_j^m}$ into f_{δ_j} and F_{δ_j}, respectively. This is taking into account the information that $\tau_j^l > \tau_i^{l:m}$, which implies $\delta_j^l > \delta_i^{l:m}$, by means of shifting the argument by $\delta_i^{l:m}$. Ultimately, this leads to a clipped density function requiring normalization which is depicted in Figure 3.7.

Using (3.33), (3.32) turns into:

$$\delta_i^m \cdot \frac{f_{\delta_j}\left(\delta_i^{l:m}+\delta_i^m\right)}{1-F_{\delta_j}\left(\delta_i^{l:m}+\delta_i^m\right)} = \frac{\partial c(\tau^{1:m}+\tau_i^m)}{\partial \tau_i^m} \cdot \frac{\mathrm{d}\tau_i^m\left(\delta_i^{l:m}+\delta_i^m\right)}{\mathrm{d}\delta_i^l}. \tag{3.34}$$

Note that the necessary normalizations of density and distribution function in (3.34) neutralize themselves. The integration of (3.34) with respect to δ_i^m and rearrangement with respect to Assumption 3.12 yields (3.24b).

To conclude, the player adapts her or his strategy in every stage if she or he has lost in the previous stage. The information of $\delta_j^l > \delta_i^{l:m}$ is used to adapt the corresponding density function of the other player for stage m, see Figure 3.7 and argument shift in (3.24b). □

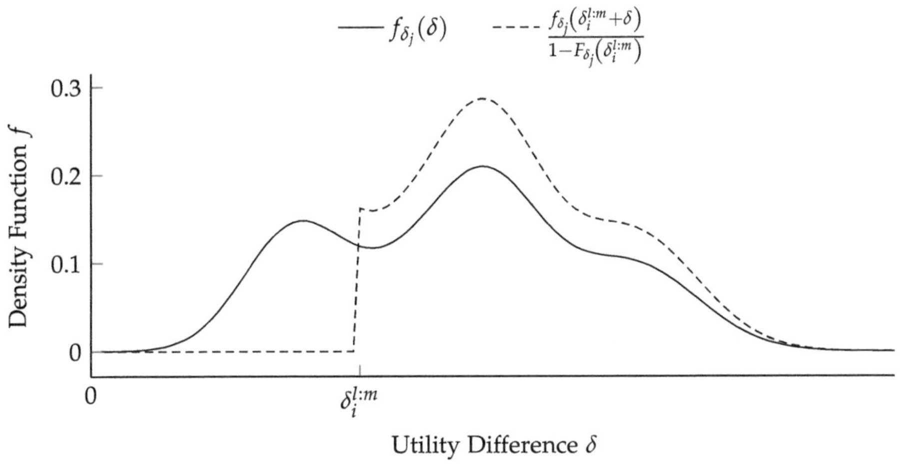

Figure 3.7: Transformation including normalization of an exemplary density function for taking $\delta_j^l > \delta_i^{l:m}$ into account.

Remark. *Both functions* (3.24a) *and* (3.24b) *fulfill Assumption 3.15 which is therefore justified: Both functions are differentiable yielding an integrable derivative and yield non-negative thresholds. The threshold functions are also invertible due to an invertible cost function (see Assumption 3.12) and a positive and diverging integrand (see [Rin14, p. 12]) resulting in a strictly increasing and therefore invertible integral.*

Note. $\tau_i^m + \tau^{l:m} = \tau_i^l$ *holds, i. e. the winning player sticks to the strategy of stage l.*

Remark. *Transformations* (3.28) *and* (3.33) *resemble the inverted threshold function which in turn depends on the cost function. The fact that these transformations are applied to threshold values of the other player j are another practical reason why Assumption 3.12 does not consider individual cost functions for both players.*

After introducing the threshold functions for all stages in Theorem 3.3, they serve as the basis of the solution strategy for the n-stage war of attrition and the following symmetric strategy profile can be defined.

Definition 3.16 (n-Stage War of Attrition Strategy Profile)
For all players $i \in P$:

1. *Start the first stage of the game with offering your preferred decision option with the highest utility.*
2. *If the other player does give in before your current threshold is reached enter the next stage by updating your threshold according to* (3.24a).
3. *If your current threshold is reached give in and start the next stage by offering the decision option with the next highest utility. Update your belief about the other player and determine the new threshold according to* (3.24b).
4. *If no agreement is reached, repeat starting with step 2.*

In the following, it is shown that this strategy profile leads to a unique perfect Bayesian equilibrium. Hence, the considered strategy profile leaves no ambiguity while following the strategies which would be present if multiple equilibria existed. This is beneficial for a practical application of the strategy profile as the uniqueness of the equilibrium does not leave open the question on which equilibrium to strive towards.

Equilibrium

In the following, it is proven that the symmetric strategy profile of Definition 3.16 leads to a unique perfect Bayesian equilibrium as stated in the following theorem:

Theorem 3.4 (Perfect Bayesian equilibrium)
Let Assumptions 3.12, 3.14 and 3.15 hold such that the symmetric strategy profile of Definition 3.16 exists. The symmetric strategy profile of Definition 3.16 yields a unique perfect Bayesian equilibrium.

Proof:
The perfect Bayesian equilibrium is defined as a refinement of the Bayesian Nash equilibrium, see Section C.1. Therefore, it has to be shown that the introduced strategy and associated beliefs fulfill the following two conditions as given in Definition C.3:

- Sequential rationality of strategies: The strategy has to be optimal with respect to the current belief of the other player's type.
- Consistency of beliefs: The belief about the other player has to be updated with respect to the observed actions of the other player.

First, it is proven that the introduced strategy is a best response to itself with respect to to the given belief about the other player's type: If both players prefer the same option at the start of the game, an agreement is reached immediately without costs. If players prefer different options, both will realize the conflict and hence the game they are in. By following the introduced symmetric strategy both players will wait until their thresholds for giving in are reached. Under Assumptions 3.12, 3.14 and 3.15, Theorem 3.3 provides that the thresholds (3.24a) and (3.24b) optimize, i. e. maximize, in expectation the individual payoff for all positive times in every individual stage $m \in \{1, \ldots, n\}$ of the game. Under Assumption 3.14, this payoff's optimality also applies for the overall game.

Second, the belief has to be updated: For this, it is referred to the proof of Theorem 3.3 which provides the necessary consideration of updating the density distribution of the utility difference in every stage with respect to the current role (winner/loser) of each player.

In summary, both conditions are fulfilled and therefore the introduced symmetric strategy profile yields a perfect Bayesian equilibrium. The uniqueness of the equilibrium follows from the deterministic relation between decision option and its utility and the deterministic calculation of thresholds in Theorem 3.3, see [FT91, p. 219],[BK99]. □

After the introduction of the n-stage war of attrition and the adaptive negotiation model, the following section highlights the models' theoretical similarities and differences.

3.4 Theoretical Comparison of the Proposed Models

Both above introduced mathematical behavior models enhance existing models to suit the scope of human-machine cooperation on decision level, see Figure 2.7. Consequently, both mathematical behavior models possess some similarities but also focus on different aspects of human-machine cooperative decision making. After a brief recapitulation of the mathematical behavior models' setup, the following paragraphs elaborate on these similarities and differences.

The adaptive negotiation model proposes a time-based concession strategy as the instant reaction strategy within negotiation and introduces the asynchronous negotiation protocol removing communication restrictions. Furthermore, the model provides an identification component based on Bayesian learning. Thereby, it addresses the identification challenge in human-machine negotiation arising due to the expected limited communication with few symbols. Upon this identification component, the model extends state-of-the-art negotiation models by an explicit adaptation strategy of negotiation behavior allowing for efficient negotiations. The adaptation strategy also yields high flexibility in modeling as it can be changed independently of the other parts of the negotiation model.

The n-stage war of attrition builds upon the conventional war of attrition game model with incomplete information and two rational players. It enhances the conventional war of attrition by allowing for more than two decision options and by a time-dependent disagreement cost function. The proposed solution strategy is proven to lead to a perfect Bayesian equilibrium.

In consequence, the proposed models fulfill the requirements and limitations stated in Section 3.1 as the adaptive negotiation model and the n-stage war of attrition consider two emancipated, equally performant, rational agents/players in a cooperative decision making scenario with multiple decision options. Agents exhibit a concessive behavior due to their lack of information on the other agent's decision option utilities and hence preferences. In other words, both above introduced mathematical behavior models of cooperative decision making represent an answer to the first research question of this thesis, see Section 2.4.

In what follows, the differences with respect to major features of the newly proposed mathematical behavior models are compared. To this end, Table 3.1 provides an overview on these features for both models. A first difference between the models is the relation between the communicated offers and the decision options: while the n-stage war of attrition requires a bijective mapping between offers and decision options, the adaptive negotiation model allows for offers conveying more information besides the proposed decision option which may be beneficial for the identification of negotiation behavior. Furthermore, the models differ in their ways of concession modeling, more specifically in the source of the decision-making pressure: the adaptive negotiation model focuses on the deadline whereas the n-stage war of attrition considers increasing time-dependent disagreement costs which may resemble a soft deadline. As a result, the adaptive negotiation model guarantees an agreement within a set period of time in contrast to the n-stage war of attrition. This difference in the agreement characteristic reflects the different origins of the two models: negotiation theory typically relies on a conflict deal in cases no agreement is found. As conflict deals cannot generally be suitably defined in the context of human-machine cooperation, this feature is implicitly integrated into the time-based concession strategy considering the deadline. On the other hand, game theory usually focuses on rational, emancipated players and hence the original war of attrition does not consider deadlines. Both models also differ in their information bases and adaptation techniques: the adaptive negotiation model allows for agents to identify the negotiation behavior of the other agent during the negotiation and to adapt their negotiation behavior over negotiation rounds. The n-stage war of attrition inherently models the uncertainty due to the incomplete information setting and takes into account each observed event in the course of the game to potentially gain and instantly utilize information about the other player. Although both models consider adaptation techniques, the general concession behavior within a single cooperative decision making process persists. Both models are also able to represent long-term adaptations, i. e. some sort of learning, of both cooperation partners. However, the adaptations' analysis is not within the scope of this thesis, see Assumption 3.7.

Table 3.1: Features of the proposed models of cooperative decision making.

Feature	Adaptive Negotiation Model	n-Stage War of Attrition
Relation: Offer to Decision Option	offers may contain more information	identical
Concession Modeling (time dependent)	target utility & (hard) deadline	disagreement costs (soft deadline)
Agreement	guaranteed	not guaranteed
Information Basis	online identification	uncertainty modeling
Adaptation	online & over rounds	online & event-based

To conclude, the adaptive negotiation model has its strengths in the ability to adapt in changing decision environments and in the agreement guarantee in highly time-sensitive situations. The latter aspect however assigns the correspondingly designed automation the feature to ultimately concede which a human decision maker could presumably take advantage of. In contrast to this, the n-stage war of attrition model has its strengths in capturing more egoistic, human traits and will yield a less concessive automation, potentially displaying stubborn behavior. Furthermore, the n-stage war of attrition model only allows for the implementation of soft deadlines and is therefore not suitable for highly time-sensitive situations. Apart from this, the n-stage war of attrition model focuses on the uncertainty of decision making scenarios and is therefore predesignated for corresponding implementations.

4 Towards the Application of Models

Subsequent to the theoretical introduction of the two mathematical behavior models of human-machine cooperative decision making, i. e. the adaptive negotiation model in Section 3.2 and the n-stage war of attrition in Section 3.3, this chapter focuses on the mathematical behavior models' practical applications and strives to answer the second research question of this thesis on how to design the corresponding automation which is capable of participating in an emancipated cooperative decision making process with a human, see Section 2.4. To this end, Section 4.1 reports on a study which investigated the suitability of both mathematical behavior models to describe human concession behavior. Moreover, Section 4.2 discusses important aspects of the model-based automation design to successfully enable the machine to cooperatively make decisions with a human.

4.1 Study on Models' Suitability to Describe Human Concession Behavior

In the following, a suitability study on the introduced mathematical behavior models of human-machine cooperative decision making of Sections 3.2 and 3.3 is presented. The study was conducted in the course of a master thesis [Wör20] and led to a publication [RWIH20]. The study investigated the mathematical behavior models' suitability to represent human concession behavior in cooperative decision making, see Section 3.1. To this end, two human participants were supposed to be confronted with a series of cooperative decision making scenarios in the original study design. However, at the time of the study it was impossible to conduct this study as planned with several participants being simultaneously in one room.[19] Therefore, a program and corresponding guidelines were designed to allow participants to conduct the study alone: the program comprised an automation capable of actively participating in cooperative decision making and provided a series of cooperative decision making scenarios to the participants by means of a graphical representation. The distribution of the program and guidelines and the collection of log-file data was conducted via email. The following sections provide information about the study's design, the results and their discussion.

---

[19] The study took place in early summer of 2020 at the height of the COVID19 pandemic. Due to imposed restrictions in Germany, it was not allowed to conduct studies with multiple participants and instructors in the same room.

4.1.1 Study Design

Based on the study's objective to examine human cooperative decision behavior, a program was implemented which displayed a series of cooperative decision making scenarios to the participants. Each scenario consisted of four decision options represented by buttons. Each decision option was associated with a different utility value visualized numerically on the corresponding button. The participants' objective was to maximize the utility values received within each and over all cooperative decision making scenarios. To this end, participants were able to select a decision option via a click on the corresponding button. The choice was visualized by a change in background color of the respective button. However, participants were not able to withdraw a choice. A designed automation acted similarly but on the basis of different utilities associated with the decision options. This intentionally caused potential conflicts on the choice of decision options. Furthermore, the participant was only able to collect utility values if she or he and the automation found an agreement on one decision option within a fixed limited time period before the next scenario began. As a result, concessive actions of the participants were expected, i. e. additional choices of decision options with decreasing utility over time. To emulate a similar behavior, the automation was programmed to also display various concession behaviors. The offers of decision options and their timestamps were recorded and fitted to simulated outcomes of the proposed cooperative decision making models to evaluate their ability to replicate human concession behavior in cooperative decision making scenarios.

In the following, the scenario setup for cooperative decision making, the decision interface (i. e. the program) and the automation behavior in the cooperative decision making scenarios is introduced in more detail. Furthermore, the study's procedure and its measures are explained.

Cooperative Decision Making Scenario

In each cooperative decision making scenario, the participant was introduced to four decision options d^{μ}, $\mu \in [1,4] \subset \mathbb{N}$ with different predefined utilities $u^{\mu}_{\mathcal{H}}$ in the range from one to seven ($u^{\mu} \in [1,7] \subset \mathbb{N}$). The range and size of both sets were chosen with the goal to not mentally overload the participants, see Section 3.1.2 and esp. Assumption 3.1. Each scenario comprised a cooperative decision making time period of $\mathcal{T} = 12\,\mathrm{s}$. This time period was based on the following motivation: Gold et al. [GDLB13] found human reaction times for driving related tasks, e. g. perceiving a hazardous situation and reacting by breaking, of around 3 s. To allow the participant to virtually perceive and react to each individual decision option, this reaction time was multiplied by four, i. e. the number of decision options within one scenario.

The participants were generally able to freely choose, i. e. offer, decision options. However, participants were not able to take back an offer they had already chosen.

The objective of the participants was to receive as much utility payoff in each scenario as possible and accumulate as much as possible throughout the series of scenarios. However, there was only a utility payoff at the end of each scenario if the participant and the automation had reached an agreement on a decision option within the given time period of cooperative decision making. In order to reach an agreement within that time period, the participants were able to concede by proposing additional decision options after their initial choice of a decision option. Therefore, a theoretical maximum of three concession steps was possible for each participant in each scenario. Due to the participants' objective, it was assumed that participants initially chose the option with the highest utility payoff and successively proposed additional decision options with decreasing utility payoff.

The automation chose decision options in a changing but predefined pattern that will be explained later. The choice of the automation was displayed to the participant. As soon as either the human or the automation chose an option that had been already offered by the other one, an agreement was reached yielding a corresponding payoff for the participant. The scenario ended if either the deadline was reached or an agreement was found.

One part of the study also investigated whether or not participants would make use of a richer communication within the cooperative decision making process. To this end, offers were not only associated with a decision option but also comprised further meaningful information for the cooperative decision making process: participants and the automation were also able to communicate the *importance level* ζ of their currently chosen decision option by double and triple clicks on that option. However, double and triple clicks reduced the potential payoff by one and two, respectively, accounting for the higher communication effort and an evaluable meaning.

Decision Interface

The decision interface of the study is depicted in Figure 4.1 by means of two exemplary screenshots. Each decision option was visualized by a button that was initially colored in light blue and had a certain utility $u_{\mathcal{H}}^{\mu}$ depicted in its lower right corner. The choice of the automation was indicated by the coloring of the respective button in dark blue. The participant was able to choose decision options by clicking on the corresponding button which then changed its color to orange. If available, the communicated higher importance level of a decision option was indicated by two or three yellow bars in the upper right corner of the decision option. If an agreement was reached, the mutually chosen decision option button turned green and the corresponding utility was added to a utility counter in the lower right corner of the screen. During the whole scenario the remaining time until the deadline had been reached was indicated by a decreasing red bar graph (i. e. inverted progress bar) in the upper half of the screen. When the scenario ended, either by reaching the deadline or an agreement, the results of this scenario were displayed for 2 s. Then, the next scenario started after a countdown of 3 s.

(a) Scenario with one offer each of participant and automation. **(b)** Scenario with communicated importance level.

Figure 4.1: Exemplary screenshots of the decision interface. ©2020 IEEE

Scenario Design

Each scenario was determined by a set of utilities for the participant ($u_{\mathcal{H}}$) and the automation ($u_{\mathcal{A}}$). However, both were unaware of each other's utilities. For each scenario, the utility patterns were assigned to decision options, i. e. the pair of utility $\left(u^{\mu}_{\mathcal{H}}, u^{\mu}_{\mathcal{A}}\right)$, $\mu \in [1,4] \subset \mathbb{N}$, was assigned to decision option d^{μ}. The decision options were presented in a random order on screen (see Figure 4.1) in order to avoid learning effects. The applied utility patterns forming different scenarios are presented in Table 4.1. The utility patterns were designed to reveal different manifestations of participants' time-based concession behaviors, which is explained in the following.

Table 4.1: Scenario utility pattern.

Scenario	$u_{\mathcal{H}}$	$u_{\mathcal{A}}$
S1	[7,5,3,1]	[1,3,5,7]
S2	[7,3,2,1]	[1,3,5,7]
S3	[7,5,3,1]	[1,2,3,7]
S4	[7,6,2,1]	[1,3,5,7]
S5	[7,5,3,1]	[1,2,6,7]
S6	[7,5,3,2]	[3,5,7,2]
S7	[7,5,3,1]	[–,–,–,7]
S8	[7,5,3,1]	[–,–,6,7]
S9	[7,3,2,1]	[–,–,6,7]

Scenario S1 had a linear utility distribution for both participant and automation. In scenario S2 and S4, the automation had a linear utility distribution and the participant faced a larger utility gap between highest and second-highest valued options and between second highest and third-highest valued options, respectively. In scenario S3 and S5, this was set vice-versa for participant and automation. All scenarios mentioned up to now let to an agreement after a maximum of three concession steps in total. In contrast to that, scenario S6 had a decision option that was the least valued option for both decision makers, i. e. at maximum two concession steps by any decision maker were required to find an agreement. Scenarios S7 to S9 caused a stubborn behavior of the automation (options with "–" were treated as not existing) to avoid the impression that the automation was forced to reach an agreement and to incite more offers of the participants within one scenario.

Automation Design

The behavior of the automation was predefined with respect to the utility pattern of Table 4.1 and the basic negotiation model for human-machine cooperation introduced in Section 3.2.3. This model was chosen without any explicit knowledge on human conceding behavior and represents the simplest form of automation design that allows for rational and active participation in cooperative decision making.

The automation always offered the option with the highest utility ($\max_\nu u^\nu_\mathcal{A}$) at the beginning of each decision making scenario. Additional offers were placed if the linear-over-time decreasing target utility $u_{\mathrm{t},\mathcal{A}}$ became smaller than a utility $u^\mu_\mathcal{A}$ of a non-chosen decision option d^μ. Therefore, the following condition was continuously evaluated for the utilities $u^\mu_\mathcal{A}$ of all so far non-chosen decision options d^μ:

$$u^\mu_\mathcal{A} > u_{\mathrm{t},\mathcal{A}} := \max_\nu \{u^\nu_\mathcal{A}\} - \left(\max_\nu \{u^\nu_\mathcal{A}\} - \min_\nu \{u^\nu_\mathcal{A}\}\right) \cdot t/\mathcal{T} \tag{4.1}$$

with $t \in [0, \mathcal{T}]$. If applicable, the automation also communicated the importance level of its choice of decision option.[20] The corresponding times were determined analogously to (4.1) by replacing $u^\mu_\mathcal{A}$ on the left-hand side of the inequality with $u^\mu_\mathcal{A} - 1$ or $u^\mu_\mathcal{A} - 2$ for the currently chosen decision option d^μ. Note that in certain cases the utility of another decision option d^ν ($\nu \neq \mu$) was equal or greater than this reduced utility ($u^\nu_\mathcal{A} \geq u^\mu_\mathcal{A} - 1$ or $u^\nu_\mathcal{A} \geq u^\mu_\mathcal{A} - 2$). In this case, this decision option d^ν was offered instead of communicating higher importance levels.

By means of this automation design based on the basic negotiation model and scenario design, the participants faced a cooperative decision making counterpart that was rational but from their perspective unpredictable in terms of decision options preference sequence and concession behavior. Furthermore, the automation design was kept as simple as possible to minimize its influence on human behavior. This effort was made to present a human-like cooperation partner to the participants to

[20] Further information on the enriching of offers with importance information can be found in the adaptive negotiation model example in Appendix B.

get as close as possible to the original study design in which two human participants were supposed to cooperatively decide.

Procedure

Provided with the designed program and guidelines explaining the study, participants were able to conduct the study on their own. Hence, participants received the program and guideline via email after they reacted to the study's invitation. In the following, the different parts of the study and their sequence are presented. The accomplishment of all practical parts of the study took up to 15 min.

1) **Information & Preparation**
   Firstly, the participants were instructed to read the guidelines on how to conduct the study. These included a user guide for the program and an explanation of which information was needed to be sent back to the examiners. Furthermore, the participants were informed about the setup of the decision scenarios (four decision options, deadline, automation also places offers, payoff only in case of agreement) and what their objective was (accumulate as much utility as possible). They were unaware of the exact behavior of the automation. Finally, they were asked to start the program.

2) **First Trial Part**
   This part of the study was a random series of scenario S1 to S8. To get to know the general handling of the program it was possible to repeat this part any number of times. The results of this part were not included in the evaluation.

3) **First Test Part**
   This part comprised three times scenarios S1 to S7, twice scenario S9 and once scenario S8 in random sequence.

4) **Second Trial Part**
   This part was built similar to the first trial part. The ability to communicate the importance level of a decision option's choice via double and triple clicks was available in this part and was the only difference regarding the usability. Furthermore, this part was not repeatable.

5) **Second Test Part**
   This part had a setup equivalent to the first test part while the ability to communicate the importance level of decision option's choice was given.

6) **Postprocessing**
   The participants were asked to send back the log-files created by the program along with additional information about age, sex and profession.

General Evaluation Procedure

The resulting data of each participant were the placed offers of each scenario, i. e. decision options that were chosen with a certain amount of clicks, and the corresponding time stamp relative to the start time of each scenario.

In a first step, rationality of participants was verified by searching for guideline violations such as offering options with increasing utility over time, not reaching an agreement or only placing single offers at the beginning or end of a scenario. These behaviors do not resemble a rational cooperative decision making process. Therefore, the data of the corresponding scenario was excluded from further examination.

In a next step, the models of cooperative decision making, namely the adaptive negotiation model and the n-stage war of attrition, were fitted to the observed participants concession behavior and the fitting error was evaluated. This was possible due to the study's design that specified the available decision options and corresponding offers, their utilities and the time frame of cooperative decision making.

The specific evaluation procedures for each model of cooperative decision making are separately explained in the following.

Evaluation Procedure for the Adaptive Negotiation Model

In the case of the adaptive negotiation model, the concession behavior within each negotiation was determined by the basic negotiation model's concession strategy, see Section 3.2.3. The basic idea of this strategy is to compare utilities $u_{\mathcal{H}}^{\kappa}$ of offers o^{κ} to a time-dependent target utility $u_{\mathrm{t},\mathcal{H}}$. If $u_{\mathcal{H}}^{\kappa} > u_{\mathrm{t},\mathcal{H}}$ holds for the first time for o^{κ} then this offer is proposed and becomes part of the offer history $O_{\mathcal{H}}^{\mathrm{H}}$. A parametric description of the target utility facing a deadline at time $\mathcal{T}$ without normalization is

$$u_{\mathrm{t},\mathcal{H}}(t,\epsilon) = \max_{\kappa}\{u_{\mathcal{H}}^{\kappa}\} - \left(\max_{\kappa}\{u_{\mathcal{H}}^{\kappa}\} - \min_{\kappa}\{u_{\mathcal{H}}^{\kappa}\}\right) \cdot (t/\mathcal{T})^{1/\epsilon} \tag{4.2}$$

with the concession parameter ϵ, see Definition 3.7. Utilizing this model, the participants' negotiation behavior can be expressed by means of their concession parameters. To determine the concession parameter of one participant within one scenario, all times $\{t_{\mathcal{H}}^{\kappa}|\kappa > 0\}$ at which the participant proposed an additional offer after the initial offer were taken into account. Note that the initial value does not provide information on participants concession behavior. Therefore, participants were instructed to propose the initial offer shortly after the start of the decision scenario. By means of the following optimization of the squared error between the concession model (4.2) with respect to $\{t_{\mathcal{H}}^{\kappa}|\kappa > 0\}$ and the set of utility $\{u_{\mathcal{H}}^{\kappa}|\kappa > 0\}$ of the observed offers $\{o_{\mathcal{H}}^{\kappa}\}$ the concession rate was estimated:

$$\hat{\epsilon} := \arg\min_{\epsilon} \sum_{\kappa>0} \left(u_{\mathcal{H}}^{\kappa} - u_{\mathrm{t},\mathcal{H}}(t_{\mathcal{H}}^{\kappa},\epsilon)\right)^2 . \tag{4.3}$$

On the basis of these estimated concession parameters of every scenario and each participant, the following aspects were evaluated:

- **Suitability of Time-based Concession Model**
  For scenarios with more than two observed offers the concession rates were estimated according to (4.3) and the resulting maximum deviation between model and observations in terms of time ($\Delta_{\max}t$) and utility ($\Delta_{\max}u$) were evaluated:

$$\Delta_{\max}t := \max_{\kappa>0} \left| t^{\kappa}_{\mathcal{H}} - (u_{t,\mathcal{H}})^{-1}(u^{\kappa}_{\mathcal{H}}, \hat{\epsilon}) \right| \tag{4.4}$$

$$\Delta_{\max}u := \max_{\kappa>0} \left| u^{\kappa}_{\mathcal{H}} - u_{t,\mathcal{H}}(t^{\kappa}_{\mathcal{H}}, \hat{\epsilon}) \right| \tag{4.5}$$

- **Influences of Valuation Pattern and Automation Behavior**
  The utility pattern of scenario S1, S2 and S4 varied in the utility that was displayed to the participant while the utility of the automation and hence its behavior stayed invariant. Therefore, these scenarios of the first test part were used for examining the influence of different utility patterns on the participants' behavior, i. e. $\hat{\epsilon}$. This was conducted by means of a non-parametric Kruskal-Wallis test by ranks [KW52] for each participant considering these scenarios.

  The utility pattern of scenario S1, S3 and S5 varied in utility considering the automation and hence the behavior of the automation also varied while the utility for the participant did not change. A Kruskal-Wallis test by ranks was applied for each participant with respect to these scenarios of the first test part to examine if the change of automation behavior influenced the participants' behavior.

- **Influence of Richer Communication**
  The influence of richer communication, i. e. in this case the ability to show the importance of a current choice to the automation, on the negotiation behavior was examined by comparing the concession parameters of both test parts. Concession rates were estimated with respect to changes of decision options disregarding changes in importance level in order to achieve a simple and comparable evaluation. The comparison was performed by means of a Kruskal-Wallis test by ranks [KW52].

Evaluation Procedure for the n-Stage War of Attrition

In case of the n-stage war of attrition, the relevant model component describing the concession behavior is the time-dependent cost function since the other model components, i. e. utility differences and their distribution, are specified by the study's design. Hence, the following measures focus on the estimation of the cost function.

Note that the n-stage war of attrition does not consider additional communication symbols like the importance of a choice. Therefore, the data of the second test part of the study associated with the ability of richer communication is not considered in this game theoretic evaluation.

For the presented evaluation, the offers, i. e. the chosen decision options and corresponding time stamps, determined the stages and corresponding thresholds of the n-stage war of attrition game model for each scenario, see Section 3.3.5: a stage is defined as the time period between two proposals of decision options by any player, i. e. cooperation partner. The thresholds describe the times after the beginning of a stage at which players concede and propose their next decision option if the other player has not yet conceded.

The evaluation was based on the postulated threshold calculation of the human player for each stage of the game according to Theorem 3.3. This calculation of thresholds $\tau_{\mathcal{H}}^{m}$ depends on the time-dependent cost function $c(t)$, the current utility difference $\delta_{\mathcal{H}}^{m}$, the corresponding utility differences density function $f_{\delta_{\mathcal{A}}}$ and on whether or not the human player has won or lost the previous stage(s) of the game. Although $f_{\delta_{\mathcal{A}}}$ was specified by the study's design, the automation in this study did not behave according to Theorem 3.3 because its behavior was governed by the basic negotiation model for reasons of simplicity. Therefore, it was assumed that the participants' beliefs of $f_{\delta_{\mathcal{A}}}$ was a uniform distribution within the given range of utility differences. Consequently, all other dependencies of the threshold calculation were known except for the cost function. In order to make the identification of cost functions manageable, an exponential function structure was assumed which yielded a parameterized cost function:

$$c(t, \boldsymbol{\theta}) = \theta_1 \cdot t^{\theta_2}, \quad \boldsymbol{\theta} = [\theta_1, \theta_2]^{\top}, \theta_1 > 0. \tag{4.6}$$

This structure was motivated by an increasing decision-making pressure over time that becomes steeper when approaching the deadline while still disagreeing. In line with Definition 3.13, the initial costs were set to zero.

For identifying the parameters $\boldsymbol{\theta}$ of this parameterized cost function, it was assumed that the sequence of offers of both participants resulting from the simulated model had to be identical to the observed sequence. Furthermore, note that the initial offers of both agents do not provide information on their concession behavior and were therefore disregarded in the identification process. Hence, the offer times $t_{\mathcal{H}}^{\kappa}$ of observed offers $o_{\mathcal{H}}^{\kappa}$ (except the initial offers, i. e. $\kappa > 0$) of the participant and those of the automation were utilized to calculate the relevant thresholds $\tau_{\mathcal{H}}^{m}$ of the participant in every stage m of each scenario. Hence, each scenario was associated with an observed set of thresholds $T_{\mathcal{H}}$. Additionally, the parameterized model yielded a similar set of thresholds $T_{\boldsymbol{\theta}}$ by means of Theorem 3.3 that depended on the parameters $\boldsymbol{\theta}$. These parameters were determined with respect to the optimal fit of the set of thresholds $T_{\boldsymbol{\theta}}$ to the set of observed thresholds $T_{\mathcal{H}}$. To this end, the following objective function based on the squared error between the sets' thresholds was set up:

$$J(\boldsymbol{\theta}) := \begin{cases} \sum\limits_{\kappa=1}^{|T_{\mathcal{H}}|} \left(\tau_{\mathcal{H}}^{\kappa} - \tau_{\boldsymbol{\theta}}^{\kappa}\right)^2 & |T_{\mathcal{H}}| = |T_{\boldsymbol{\theta}}| \\ \sum\limits_{\kappa=1}^{|T_{\mathcal{H}}|} \left(\tau_{\mathcal{H}}^{\kappa} - \tau_{\boldsymbol{\theta}}^{\kappa}\right)^2 + \sum\limits_{\kappa=|T_{\mathcal{H}}|+1}^{|T_{\boldsymbol{\theta}}|} \left(\tau_{\boldsymbol{\theta}}^{\kappa} - \mathcal{T}\right)^2 & |T_{\mathcal{H}}| < |T_{\boldsymbol{\theta}}| \\ \sum\limits_{\kappa=1}^{|T_{\boldsymbol{\theta}}|} \left(\tau_{\mathcal{H}}^{\kappa} - \tau_{\boldsymbol{\theta}}^{\kappa}\right)^2 + \sum\limits_{\kappa=1}^{|T_{\boldsymbol{\theta}}|} \left(\tau_{\boldsymbol{\theta}}^{\kappa} - 0\right)^2 & |T_{\mathcal{H}}| > |T_{\boldsymbol{\theta}}| \wedge |T_{\boldsymbol{\theta}}| \neq 0 \\ \left(c(\mathcal{T}, \boldsymbol{\theta}) - 1.5 \cdot c(\mathcal{T}, \hat{\boldsymbol{\theta}})\right)^2 & |T_{\boldsymbol{\theta}}| = 0. \end{cases} \tag{4.7}$$

The different cases with respect to $T_{\mathcal{H}}$ and $T_{\boldsymbol{\theta}}$ was utilized to ensure an identical sequence of offers, i. e. thresholds, between the observation and the simulated parameterized model. The penalty components in the cases in which the number of offers was not identical created an incentive to either reduce or increase the number of thresholds in the simulated set $T_{\boldsymbol{\theta}}$. In the case that the simulated model did not provide a single threshold $\tau_{\boldsymbol{\theta}}$, the comparison of cost function values at the end of the scenario with respect to the current parameters $\boldsymbol{\theta}$ and estimated parameters of a previous optimization iteration $\hat{\boldsymbol{\theta}}$ created an incentive to increase the values of parameters $\boldsymbol{\theta}$ and hence the cost function values. With these increased cost function values, the simulation of the model yielded thresholds $\tau_{\boldsymbol{\theta}} < \mathcal{T}$.

Minimizing the objective function (4.7) by iteratively simulating the n-stage war of attrition with respect to parameters $\boldsymbol{\theta}$ finally resulted in the identified parameters $\hat{\boldsymbol{\theta}}$ that fitted the observed thresholds to the simulated ones:

$$\hat{\boldsymbol{\theta}} = \arg\min_{\boldsymbol{\theta}} J(\boldsymbol{\theta})\,. \tag{4.8}$$

On the basis of these estimated cost function parameters for each scenario and each participant, the following aspects were evaluated:

- **Suitability of Modeling Concession by Means of a Cost Function**
  For scenarios with more than two observed offers, i. e. at least two more offers after the initial offer, the two cost function parameters could be unambiguously estimated according to (4.8) and the resulting maximum deviation between simulated and observed thresholds was calculated:

$$\Delta_{\max}\tau := \max_{\kappa>0} \left|\tau_{\mathcal{H}}^{\kappa} - \tau_{\hat{\boldsymbol{\theta}}}^{\kappa}\right| \tag{4.9}$$

- **Generalizability of the Cost Function**
  According to Definition 3.13, the cost function is supposed to be common knowledge and equal for all players. Regarding the practical application of the n-stage war of attrition, it would be beneficial if the cost function generalizes over different scenarios. Consequently, the above introduced estimation (4.8)

of individual parameters for every scenario was augmented to investigate this generalizability: three groups comprising increasing number of sets of scenarios and participant quantities were defined and different parameter sets $\hat{\theta}$ were estimated for each group. The three groups only consisted of scenarios S1 to S5 as these represent situations in which potentially both cooperation partners conceded in order to reach an agreement. The groups were defined as follows:

G1: *Parameter Determination Depending on Types of Scenarios*
For this group, a common cost function for each type of scenario and each participant was postulated. Consequently, one parameter set $\hat{\theta}$ was determined for all scenarios of one type and for each participant by means of (4.8). Hence, there were five (for scenario types S1 to S5) times the number of participants parameter sets that minimized the timely deviation between simulated and observed thresholds.

G2: *Parameter Determination Depending on all Scenarios*
Postulating, there was only one common cost function for each participant, this group comprised all scenarios of each participant. Hence, one parameter set for each participant was determined by means of (4.8) that minimized the timely deviation between simulated and observed thresholds for all scenario types of the respective participant.

G3: *Parameter Determination Depending on all Scenarios and all Participants*
Lastly, one parameter set was determined by means of (4.8) that minimized the timely deviation between simulated and observed thresholds for all scenario types and all participants.

The influence of considering these groups with respect to the parameter estimation and the corresponding timely deviations of simulated and observed thresholds was evaluated by means of the Kruskal-Wallis test [KW52].

Participants

27 participants (70.4 % male, 29.6 % female) with a range of 22 to 56 years (average age of 29.2 years) took part in the study. The majority of participants were research associates or engineers (37 %) and students (29.6 %). Participants were recruited without any intended selection procedure and compensation.

4.1.2 Results Concerning the Adaptive Negotiation Model

This section presents the results concerning the adaptive negotiation model. Due to fact that three participants violated the study's guidelines by not striving for the highest utility and therefore did not provide any information about their concession behavior, the data of 24 participants is presented in the following.

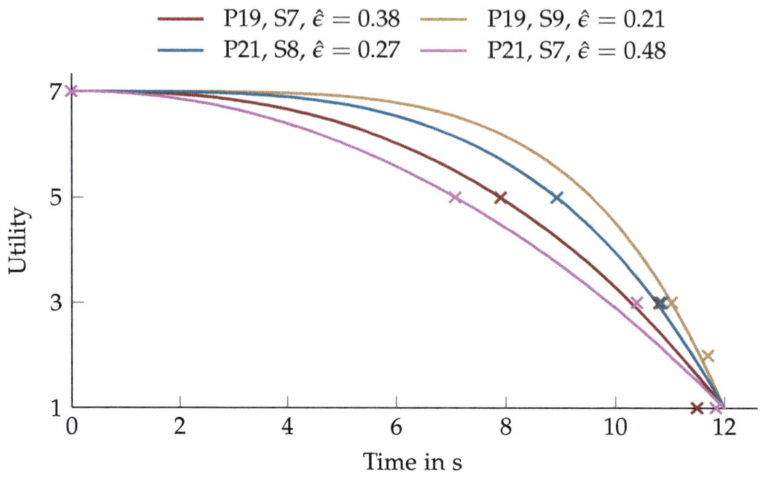

Figure 4.2: Exemplary observed offer times (×) and corresponding target utility trajectories of participants 19 and 21 in different scenarios.

Concession Model Fit of Target Utility

In order to analyze the proposed concession model based on the target utility concept, the differences between the fitted model and observed offers of the participants were calculated, see (4.3). Exemplary observed offers and corresponding identified target utility trajectories are depicted in Figure 4.2. The estimated concession rates $\hat{e}$ were in the range of 1.6×10^{-3} to 0.9 with a mean value of $M = 0.2$ and a standard deviation of $SD = 0.22$. The resulting model errors considering time ($\Delta_{\max} t$) and utility ($\Delta_{\max} u$) are presented in Table 4.2. Furthermore, Table 4.2 also provides the average (M) and the standard deviation (SD) of the maximum error based on 62 valid examinations. The deviations $\Delta_{\max} t$ and $\Delta_{\max} u$ were within the range of 3.6×10^{-3} ms to 7.3×10^{3} ms and 2.9×10^{-6} to 3.94, respectively.

Table 4.2: Exemplary, highest and average target utility model errors. Overall analysis comprised 62 scenarios with more than two observed offers of 16 participants.

Scenario	$\boldsymbol{\Delta_{\max} t}$ **in** ms	$\boldsymbol{\Delta_{\max} u}$
P19: S7	314.4	0.34
P19: S9	292.8	0.44
P21: S8	908.4	0.84
P21: S7	121.2	0.19
M	419.8	0.33
SD	978.3	0.54

Observed Decision Option Offers

In Figure 4.3 compact boxplots (see explanations in the Appendix D.1) for observed timestamps of decision option offers of all participants and scenarios S1 to S5 of the first test part are depicted. The majority of first offers was placed before 2 s. There was a large variance in time among the participants when they were about to place second and possibly third offers, e. g. participants 3, 4, 5, 11, 19, 22, 23 and 24. No correlation with the different scenarios was noticeable. Furthermore, some participants placed their second offer exclusively in the last second of the scenario, e. g. participants 1, 7, 10, 13, and 14.

These observations are also visible in Figure 4.4 presenting the identified concession rates $\hat{e}$ for each individual scenario in compact boxplot manner. The concession rate was in the range of 0.001 to 2.8 and had a great variance among participants. The strategy to place the second offer in the last second of the study let to concession rates close to zero.

Influences of Valuation Pattern and Automation Behavior

In order to apply the Kruskal-Wallis test by ranks to evaluate the similarity of participants' negotiation behavior facing different utility patterns, 23 valid sets of measurements were obtained. At a significance level of 5 % 19 participants (82.6 %) did not vary their behavior with respect to facing different utilities. Four participants (17.4 %) did: participants 5, 17, 22 and 24.

Similarly, 22 valid sets of measurements were available for applying the Kruskal-Wallis test by ranks to examine the similarity of participants' negotiation behavior facing different automation behaviors. The behavior of 18 participants (81.8 %) was not significantly influenced ($\alpha = 5\,\%$), 18.2 % (four participants) were influenced by this change of automation behavior: participants 3, 5, 17 and 18.

Influences of Richer Communication

The compact boxplots of identified concession rates of scenarios S1 to S5 of both test part 1 and 2 and every participant are depicted in Figure 4.5. Applying the Kruskal-Wallis test by ranks with a significance level of $\alpha = 5\,\%$ to compare the distributions of $\hat{e}$ of both test parts yielded that 79.2 % of the participants did not adapt their negotiation behavior. Participants 1, 7, 10, 13 and 16 showed significant differences. However, six participants (25 %) did not utilize the richer communication feature, e. g. participants 6, 7 and 23. 12 participants (41.7 %) occasionally and six participants (33.3 %, e. g. participants 5, 16 and 19) intensively utilized this feature.

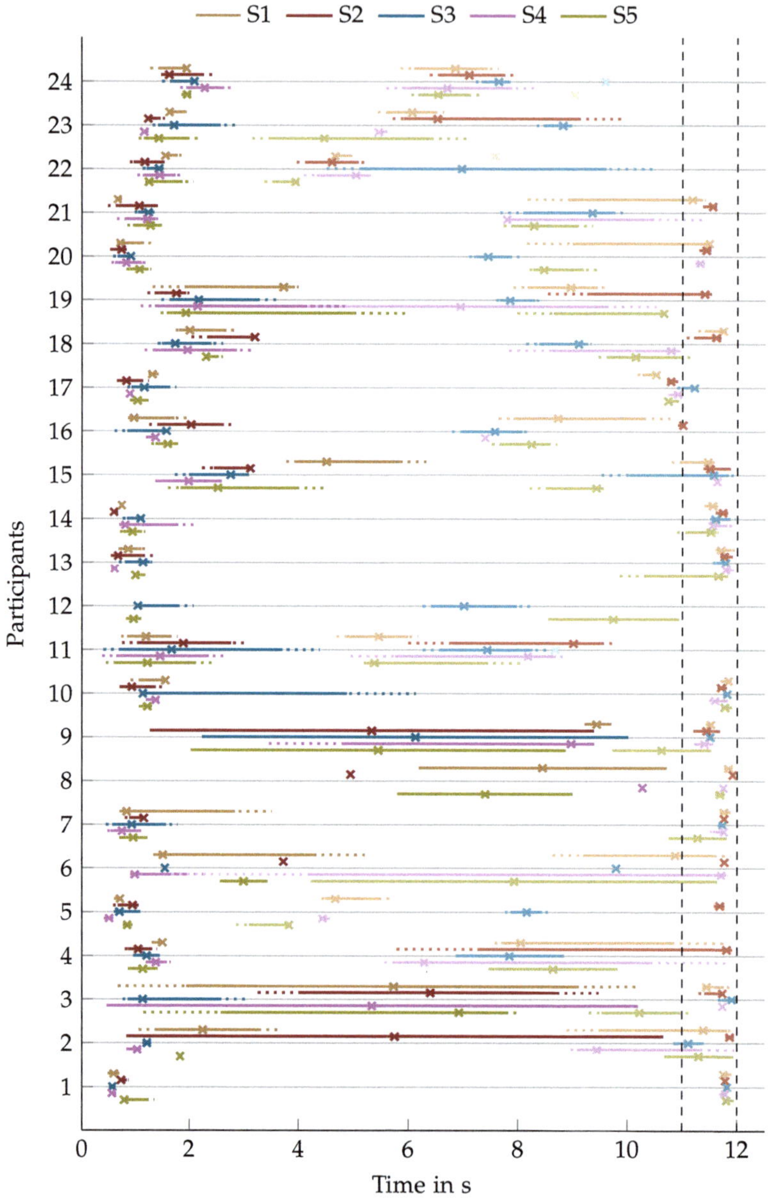

Figure 4.3: Compact boxplots (see explanation in Appendix D.1) of observed offer timestamps for each participant, individually for scenario types S1 to S5 based on data of test part 1: colors fade with number of offers. For all scenarios: median ×, lower/upper quartile –, lower/upper adjacent · · · .

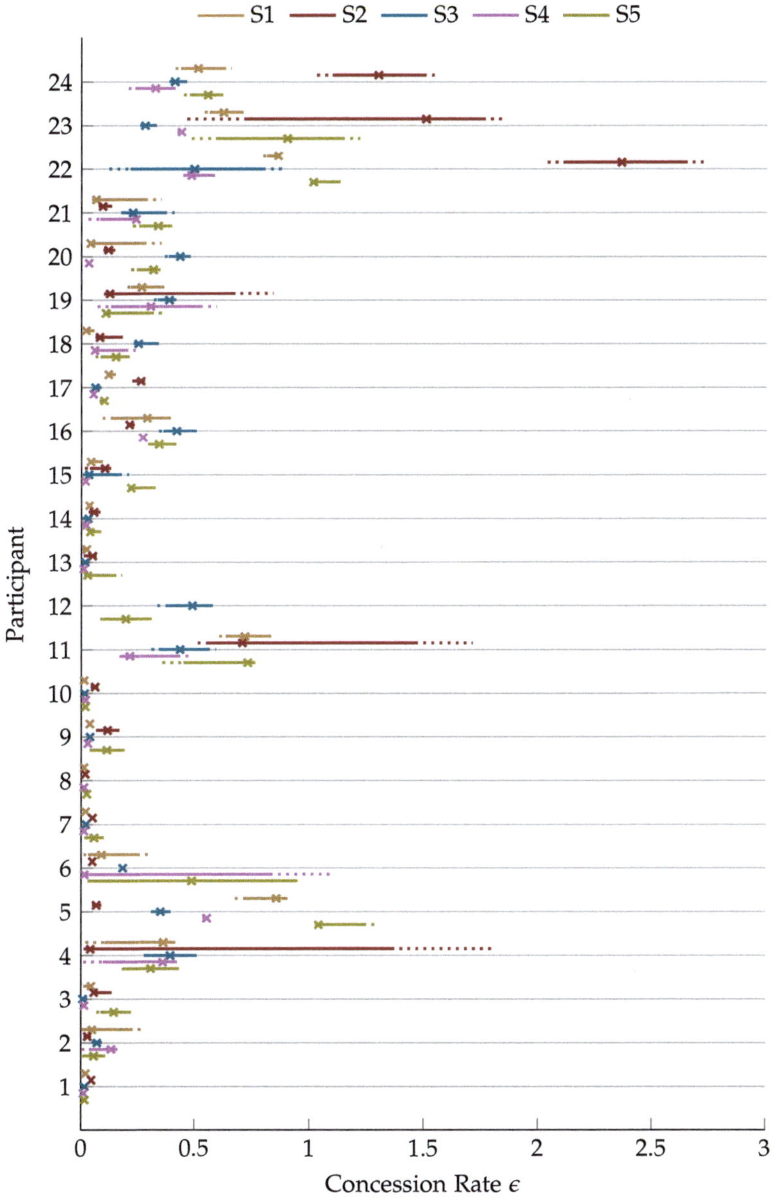

Figure 4.4: Compact boxplots (see explanation in Appendix D.1) of identified concession rates for each participant, individually for scenario types S1 to S5 based on data of test part 1. Median $\times$, lower/upper quartile –, lower/upper adjacent $\cdots$.

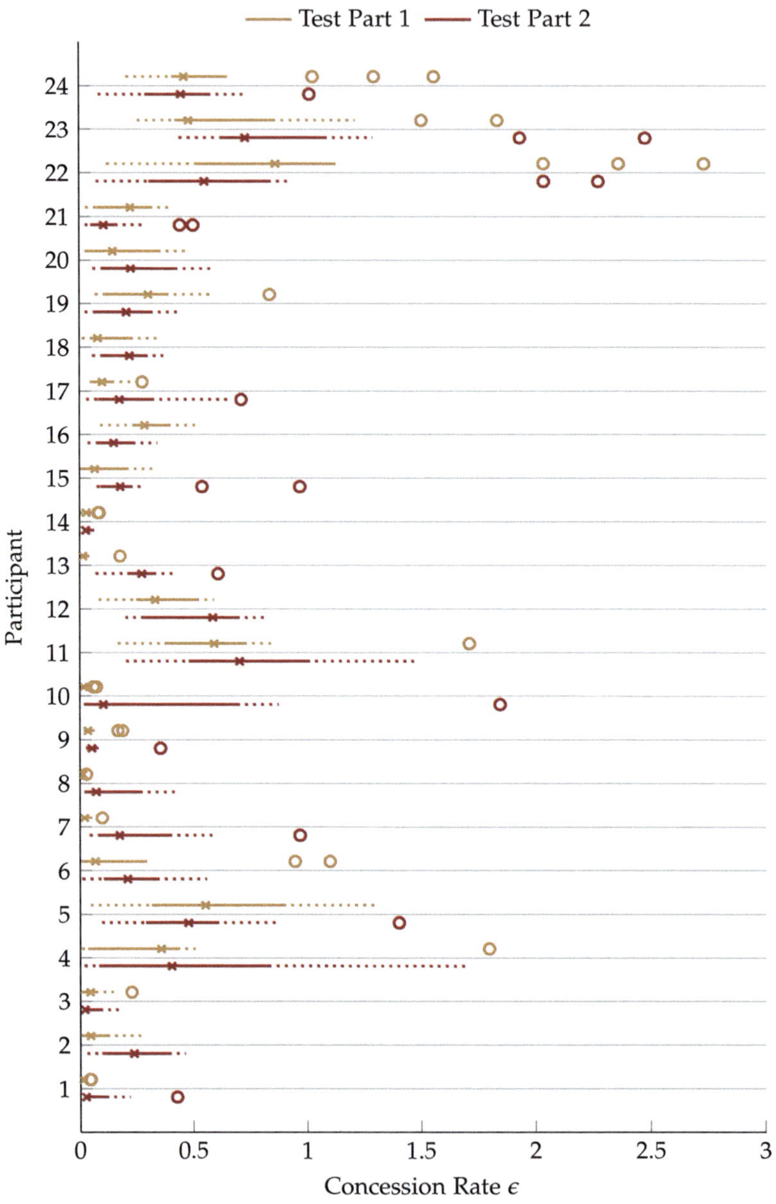

Figure 4.5: Comparison of compact boxplots of concession rates of scenarios S1 to S5 of test part 1 and 2. Median $\times$, lower/upper quartile –, lower/upper adjacent $\cdots$, outliers $\circ$.

4.1.3 Results Concerning the n-Stage War of Attrition

In this section, the results concerning the game-theoretic n-stage war of attrition model are presented. Equally to the above presentation of results concerning the adaptive negotiation model, the data of 24 participants is presented in the following.

Concession Model Fit of the Cost Function

To examine the proposed concession modeling by means of a cost function, the maximum deviation between simulated and observed thresholds was calculated for scenarios with more than two human offers. As for the target utility model examination, there were 62 of these scenarios originating from 16 participants. Figure 4.6 provides exemplary identified cost functions of four participants for one scenario each. The fitted parameters were in the range of 1.7×10^{-10} to 9.9 ($\hat{\theta}_1$) and 4.9×10^{-4} to 14.5 ($\hat{\theta}_2$) with mean values of $M = [1.06, 1.95]$. Table 4.3 provides the maximum deviation ($\Delta_{\max}\tau$) between observed thresholds and the ones corresponding to the identified cost function for the exemplary scenarios of Figure 4.6 and the mean (M) and standard deviation (SD) of all maximum deviations $\Delta_{\max}\tau$ from all applicable 62 scenarios. The deviations $\Delta_{\max}\tau$ were within the range of $4.4 \cdot 10^{-5}$ ms to 4813 ms.

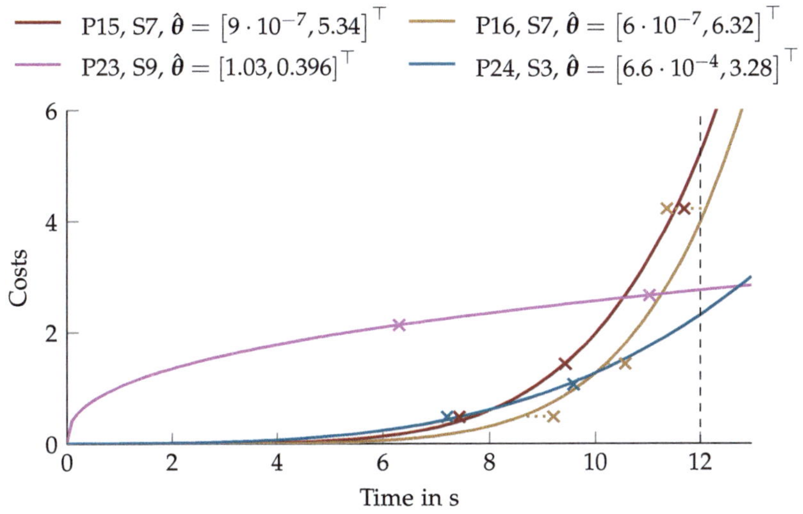

Figure 4.6: Exemplarily identified cost functions based on observed thresholds (×). The vertical dashed line visualizes scenarios' deadline at 12 s.

Table 4.3: Exemplary and average cost function model errors. Overall analysis comprised 62 scenarios with at least two observed offers of 16 participants.

Scenario	$\Delta_{\mathbf{max}}\tau$ **in** ms
P15, S7	254.51
P16, S9	745.91
P19, S7	704.87
P23, S9	$5.5 \cdot 10^{-5}$
M	416.67
SD	543.03

Generalizability of the Cost Function

Figure 4.7 shows the compact boxplots of the maximum deviations of observed and simulated thresholds depending on the defined scenario groups G1 to G3 for each participant and scenario type S1-S5 separately. Table 4.4 provides the maximum and average deviation between observed and simulated thresholds based on the identified parameters considering different scenario groups G1 to G3 of scenario types S1 to S5. The statistical analysis by means of the Kruskal-Wallis test by ranks yielded a significant difference of deviations of observed and simulated thresholds with respect to scenario groups G1 to G3. A pairwise post-t-test revealed that the distribution for G1 was significantly different compared to G2 and G3, whereas there is no significant difference between G2 and G3.

Table 4.4: Average and maximum deviation between observed and simulated thresholds depending on scenario groups.

	$\Delta_{\mathbf{max}}\tau$ **in** ms		
	G1	**G2**	**G3**
M	671.4	3587.1	4368.5
SD	905.6	2618.5	2802.3
max	6735.0	9287.6	9000.5

4.1.4 Discussion

In general, participants displayed a diverse concession behavior regarding observed times of decision option offers as depicted in Figure 4.3. However, no distinct influence of the scenario types differing in the utility patterns was noticeable. The initial decision option choice was usually offered within two seconds. This reflects human reaction time for consciously conducted tasks (about 2 s, see [GDLB13]). However, considering the countdown phase before each cooperative decision scenario and the

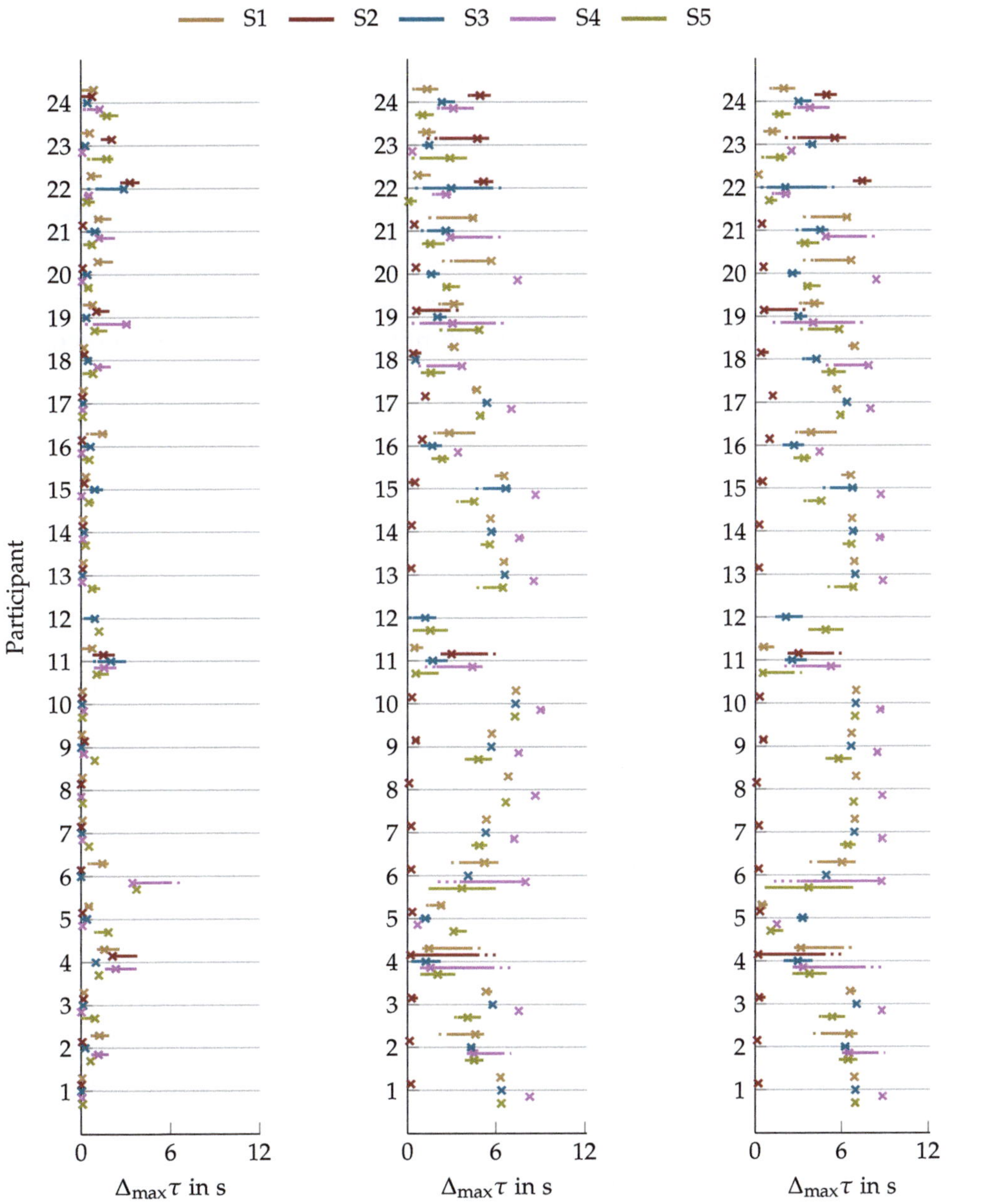

(a) G1: parameters optimized separately for each scenario type and participant

(b) G2: parameters optimized for all scenarios of each participants

(c) G3: parameters optimized for all scenarios of all participants

Figure 4.7: Compact boxplots of maximum deviations $\Delta_{max}\tau$ between observed and simulated thresholds for scenario groups G1 to G3 provided for each participant and scenario type S1-S5 separately based on data of test part 1. Median $\times$, lower/upper quartile –, lower/upper whisker $\cdots$, outlier $\circ$.

rather small number of decision options and utilities, the reaction times appear rather long. This may have been influenced by the interface design or the rapid sequence of decision options within the study. Hence, cooperative decision interface design needs to ensure the mutual start of the cooperative decision making scenario. Furthermore, some participants exhibited a two-offers-strategy, i. e. participants placed their second offers only close to the deadline without any noticeable dependence of utilities on the concession behavior. Hence, the study design did not encourage all participants to consciously evaluate utilities and choose a corresponding concession strategy leading to a cooperative decision making *process*. This also applies for the participants who did not strive for the highest utility and who were therefore excluded from the evaluation because they did not provide any concession strategy information. These observed behaviors highlight the importance of elaborated study and interface designs for cooperative decision making. Some participants provided oral feedback saying that the given time made it possible to consciously decide. This demonstrates that the provided time for cooperative decision making was appropriate for the given scenario and interface design. Furthermore, this encourages the application of the same design principle (3 s times the number of decision options) in related cooperative decision making scenarios with similar interface designs.

However, for those participants who did engage in the cooperative decision making process, the basic negotiation model fit to the observed human behavior revealed the target utility model's suitability to model concession behavior. The maximum timely deviations $\Delta_{\mathrm{max}}t$ were mostly within the range of human reaction time [GDLB13]. Therefore, they can be considered to be noise caused by human actions when using the interface. When fitting the basic negotiation model, the majority of identified parameters depicted in Figures 4.4 and 4.5 was below $\epsilon < 1$. Therefore, the corresponding human negotiation behavior is considered to be "competitive" [VKG14]. This supports findings of earlier investigations of human concession rates [VKG14]. The subsequent statistical analysis of the identified concession rates yielded the insight that the concession behavior of some participants depended on the scenario types which differ in terms of utilities and automation behavior as well as on the form of communication. Furthermore, the high diversity of identified parameters of the modeled concession behavior among participants supports the general impression based on the observed times of decision option offers depicted in Figure 4.3. Consequently, an identification and adaptation functionality as provided by the adaptive negotiation model may be beneficial for the design of an automation enabled to negotiate with and adapt to the concession behavior of humans. This adaptation also has the potential to counteract the observed two-offers-strategy or other stubborn behavior: the automation may adapt either to equally stubborn negotiation behavior or to early-conceding behavior to avoid fruitless negotiations. The fact that only one third of participants intensively utilized the richer communication ability to additionally indicate the importance level of a choice shows that the other participants did not see the necessity or benefits of this form of richer communication. Hence, if some form of richer communication is applied in future, the necessity and benefit of it has to be made more apparent to the participants, including a revision of the interface design.

Regarding the n-stage war of attrition fit to the the observed human behavior, the results yielded average model errors that were also within the range of human reaction time. Hence, the errors can also be considered noise of human actions to operate the interface. Therefore, also the n-stage war of attrition can be considered a suitable model for human concession behavior in cooperative decision scenarios. However, as the examinations of the cost functions' generalizability shows, the model errors increased greatly when attempting to generalize over scenario types and over participants. Hence, although the n-stage war of attrition explicitly relies on uncertain information of the cooperation partner, in terms of automation design it may be beneficial to have some sort of adaptation technique in place to adapt to individual human behavior.

4.1.5 Conclusion

The fit of the proposed mathematical behavior models of human-machine cooperative decision making, i. e. the basic negotiation model and the n-stage war of attrition, to the observed human behavior revealed the models' suitability to model human concession behavior in cooperative decision making scenarios. Hence, the proposed mathematical behavior models are a suitable basis for the design of an automation capable to actively take part in human-machine cooperative decision making exhibiting human-like concession behavior.

The study also provided useful insights that need to be considered in the automation design based on the proposed models of cooperative decision making: The automation should be capable to adapt to individual human behavior. Furthermore, the interface design for cooperative decision making requires particular attention to ensure an intuitive and proper interaction process.

In terms of future experiments on human-machine cooperative decision making, the study showed that an intuitive interface and careful scenario design in terms of presenting decision options' utilities is crucial to encourage humans to properly perceive, comprehend and consciously choose from available decision options. Furthermore, this study forms the foundation of future experimental investigations of automation designs based on the proposed mathematical behavior models and their suitability of describing human concession behavior.

In essence, the conducted study on the models' suitability for describing human concession behavior provided the following key insights.

- The basic negotiation model and the n-stage war of attrition are suitable to describe human concession behavior.

- There are indications that an adaptive automation towards individual human concession behavior may be beneficial.
- Adequate interface designs for cooperative decision making are crucial to ensure an intuitive and fruitful interaction.

After proposing the mathematical behavior models of cooperative decision making and assessing their suitability for describing human-like concession behavior, the following section introduces the automation design for human-machine cooperation on decision level based on these mathematical behavior models.

4.2 Model-Based Automation Design

After introducing two mathematical behavior models of human-machine cooperative decision making in Chapter 3 and evaluating their suitability to represent human time-dependent concession behavior in Section 4.1, the following section describes the automation design based on these mathematical behavior models and on some general aspects of human-machine cooperative decision making. The objective of the proposed automation designs is to enable humans to establish a *mental model* of the automation's behavior. This is assumed to yield high user acceptance [FSKL08]. To facilitate the human establishment of mental models, the proposed automation designs utilize the cooperative decision making models which are capable of representing human behavior in a cooperative setting, see Section 4.1. Previous success of similar design approaches for driver assistance systems in the context of human-machine cooperation on action level [Lan02, Fla19] supports this model-based approach.

The following section discusses general aspects of automation design for cooperative decision making. Subsequent sections provide the model-specific guidelines for implementing the corresponding automation designs.

4.2.1 General Automation Design for Cooperative Decision Making

In order to design an automation which is able to take part in a cooperative decision making process, not only the the automation behavior requires attention. Also the decision making interface and the situation in which a cooperative decision making process can take place have to be considered.

Decision Options and Their Evaluation

The set of decision options needs to be defined appropriately. This includes that all decision options need be apparent and valid for both cooperation partners. In a practical implementation suitable for many areas of application, ensuring this requirement is a challenging task. Furthermore, the number of decision options within one scenario of cooperative decision making should either be limited or suitably aggregated by means of abstraction (see example in [KFS+12]) such that the human cooperation partner is able to conceive all decision options and their impact. An appropriate number may be limited to four decision options as this is the "capacity limit [of human] focus of attention at one time" [Cow01].

Additionally, there has to be at least one measure which allows for a differentiation of the decision options by both cooperation partners. Despite this, the measures do not have to be identical for human and machine. However, it may be beneficial for a fruitful human-machine cooperation if some identical aspects of the decision making scenario are considered by the measures of human and machine such that both cooperation partners' decisions are to some extent meaningful to the other partner. This aspect is crucial for the identification of and adaptation to human behavior within the cooperative decision making process.

Start, Duration and End of the Cooperative Decision Making Process

The scenario for cooperative decision making should allow for a time span in which a cooperative *process* can take place, i. e. after the initial decision making of both cooperation partners resulting in a conflict situation, there has to be time for both cooperation partners to evaluate the choice of their partner, reflect on their decisions and potentially concede by proposing different decision options. An appropriate time span obviously depends on participants' cognitive capabilities, the decision making scenario and its complexity, e. g. its number of decision options. As a consequence, cooperative decision making requires in general some time in the magnitude of human reasoning and reaction times. Therefore, it is not suitable for highly time-critical scenarios.

For practical implementations however, it is suitable to limit the time period of cooperative decision making in order to avoid confusion about the beginning of the process and to prevent an endless process without reaching an agreement. In case of defining the beginning of a cooperative decision making process, there are two potential design options assuming both cooperation partners are able to perceive the decision scenario and initially decide: from the perspective of automation design, the process may either start as soon as the automation is able to decide on its initial decision option or as soon as the human communicates her or his initial choice of decision option. In the course of the study on the models' suitability reported on in Section 4.1, the first design option often induced a purely reactive human behavior, i. e. participants would only react to choices proposed by the automation shortly before the deadline was reached. Hence, no real process of cooperative decision making was established. Therefore, it is recommended to enforce the second design option e. g. via the interface design. Besides the beginning, the end of the cooperative decision making process, i. e. the point in time until which an agreement should be reached, also requires attention. It must be designed in such a way that the overall process duration is reasonably short to come to an agreement in a timely manner, but still long enough to allow for at least skill-based or knowledge-based human action, see Section 2.2.3 and [Ras83]. In the course of the study on the models' suitability (see Section 4.1), the broad rule to set the overall time period to 3 s times the number of decision options has proven to be appropriate. This rule is based on the typical human reaction time of 3 s in driving related tasks, e. g. perceiving and reaction to driving situations, found by Gold et al. [GDLB13]. Upon this, it is proposed to virtually provide this time to perceive and react for each available decision option. However, the consequence of setting a hard deadline and potentially enforcing it via the decision making interface requires the allocation of ultimate authority in case the applied model for cooperative decision making does not guarantee to find an agreement before the deadline is reached. Depending on the area of application and the type of decision to be taken in the course of the cooperative decision making process, different allocation strategies can be utilized: if the cooperative decision making is about actions with serious influences, regulatory and ethical reasons allow only the human to be the ultimate decision maker [FDM+20]. In case the cooperative decision making is only concerned with comfort functionality, it is reasonable to also consider the automation to be the ultimate decision maker.

Decision Making Interface

As already mentioned, the decision interface between human and automation plays a key role in the general automation design for cooperative decision making. Its design is crucial as it has to enable a period of potential cooperative decision making as well as it has to make the human aware of this period by communicating its beginning and end. Furthermore, it has to present the available decision options the latest at the beginning of a decision scenario and allow for their selection by the human. Moreover, the interface has to ensure conceding-only behavior during the cooperative decision making process.

From an ergonomic perspective, the interface design has to allow for an intuitive start of the cooperative decision making process, an intuitive communication of the process' end and an intuitive presentation and selection of decision options [BD16, WWM+19, FDM+20]. Moreover, it should provide adequate feedback on mutual agreements or ultimately valid decision options if no agreement is reached in order to increase the overall system's transparency and, by association, also human trust and acceptance.

4.2.2 Adaptive Negotiation Automation Design

The automation design based on the adaptive negotiation model introduced in Section 3.2 requires some meaningful instantiations of the general design rules from above.

Measure for Utilities

In the context of negotiation theory, offers are differentiated by means of a utility measure, see Definition 3.5 in Section 3.2.3. The definition of this utility measure (3.1a) has to take into account the decision option associated with the evaluated offer and potential additional information relevant for the cooperative decision making process. The actual measure has to yield unique and meaningful utility values. Furthermore, it is suitable to design the measure in such a way that comparison of utility values between different decision scenarios is possible. This could e. g. be achieved by normalization if the range of possible utility values is known. An exemplary utility function u_t is defined in (B.1a) in Section B.2.

Parameterization of Concession Strategy, Identification and Adaptation

Apart from the utility measure definition, the automation requires an initial set of parameters, especially in terms of the concession parameter $\epsilon_\mathcal{A}$ for the target utility function u_t defined in (3.3) and other parameters for identification and adaptation.

The study on models' suitability (see Section 4.1) provides the insight that human concession rates range between 0.0016 and 0.9073 with an average value of approximately 0.2. It is therefore sufficient to set the concession rate of the automation design to a value within this range such that the automation's behavior is perceived as being human-like. Furthermore, the average value is proposed as the initial concession rate in the automation design considering the automation's ability to adapt to individual concession behavior. In terms of identification by means of Bayesian learning, the re-initialization of 10 % of the probability mass to avoid the exclusion of individual hypotheses has proven to be appropriate, see remark in Section 3.2.4. The adaptation design parameter β required in (3.11) and the risk disposition factor r required in (3.12b) have to be within the interval $]0,1]$ (see Section 3.2.5) and can be tuned with respect to the relation of negotiation time and outcome (β: the higher the value the less important becomes negotiation time in comparison to negotiation outcome) as well as with respect to the sensitivity and speed of the adaptation itself (r: the higher the value the more sensitive and faster becomes the adaptation).

With respect to the identification and adaptation aspects of the adaptive negotiation model, also the update rates need to be set. In general, the adaptation rate must be sufficiently small compared to the rate of identification in order to only adapt on the basis of converged identification results, see Section 3.2.5. Apart from that and although the identification is possible at any time even if there is no new offer of the cooperation partner (see Section 3.2.4), experience has shown that identification updates are most effective at times where new offers are placed. Therefore, the identification rate should depend on the rate of offers of the cooperation partner. This in turn depends on the partner's concession behavior and number of potentially available offers/decision options. The more concessive the partner is and the more offers are available, the more offers of this partner will be observed within one round of negotiation. If the number of observed offers within on round is expected to be close to zero, it might also be appropriate to identify (and potentially adapt) only once after each round of negotiation. This leads to purely time-based reaction behavior during a negotiation round and a cooperation partner's behavior-depended adaptation of the automation behavior after negotiation rounds, see Sections 3.1.2 and 3.2.5.

4.2.3 The n-Stage War-of-Attrition Automation Design

The automation design based on the n-stage war of attrition introduced in Section 3.3 also requires some meaningful instantiations of the general design rules of Section 4.2.1 and additionally specific considerations concerning the war of attrition model.

Defining Utility Differences and Corresponding Distributions

A meaningful utility difference measure for evaluating the decision options is required, see Definition 3.13. An appropriate utility definition as described above in Section 4.2.2 is a suitable basis which only needs customization by sorting the utilities in descending order and calculating the differences between neighboring utilities. This also yields the preference order of the decision options.

Closely related to the definitions of utilities and the corresponding differences among them is the determination of the distribution of utility differences of the human which is required for the threshold functions (3.24a) and (3.24b). Definition 3.13 of the applied war of attrition game model assumes that the distributions of utility differences is common knowledge. However, the utility difference distribution $f_{\delta_{\mathcal{H}}}$ of the human in practice is unknown to the automation. Therefore, it is proposed that the automation initially assumes a suitable distribution of utility differences of the human and adapts this distribution by means of an identification approach while engaging in the cooperative decision making process with the human. In terms of the initial assumption on the utility difference distribution $f_{\delta_{\mathcal{H}}}$, two different assumptions are proposed:

1) **Uniform Distribution**
   Assume a uniform distribution over a suitable interval of utility differences, i. e. within the same range as those of the automation. This yields a distribution requiring the least information about human utility differences.

2) **Distribution of the Automation**
   Assume that the measures of utility are similar for human and automation and hence adopt the distribution of utility differences of the automation, i. e. set $f_{\delta_{\mathcal{H}}} := f_{\delta_{\mathcal{A}}}$.

Remark. *In order for Assumption 3.15 to hold in practice, the threshold functions* (3.24a) *and* (3.24b) *require a positive and diverging integrand (see [Rin14, p. 12]) such that the threshold functions are strictly increasing and hence yield unambiguous thresholds. This has to be ensured for the assumed (or identified) density distribution of human utility differences. Special consideration is required if the calculations are discretized.*

The initially assumed distribution of the human can be updated by means of observations in the course of the game. This requires to solve the inverse game of the n-stage war of attrition, i. e. to determine utility differences δ_j^m corresponding to observed thresholds τ_j^m while taking into account the strategy determination of Theorem 3.3. One possible realization to solve this inverse game is the iterative distribution identification algorithm which is introduced in the following.

Iterative Utility Difference Distribution Identification Algorithm

This identification method was published in [RTIH20] and for reasons of applicability but without loss of generality, it is introduced here for the case in which the automation, denoted as player $\mathcal{A}$, identifies the utility difference density distribution of of the human, denoted as player $\mathcal{H}$. Hence, $f_{\delta_{\mathcal{H}}}$ is the subject of identification for player $\mathcal{A}$, i. e. the automation.

Note that $f_{\delta_{\mathcal{A}}}$ and costs $c(t)$ are common knowledge and that the concessive behavior of either player is observable by the other player. The main assumption for the iterative identification algorithm is the following:

Assumption 4.1. *Both players play the n-stage war of attrition in a perfect Bayesian equilibrium.*

Hence, both players are aware of how they determine their thresholds, see Theorem 3.3 and 3.4. Therefore, player $\mathcal{A}$ is able to use (3.24a) and (3.24b) to uniquely calculate utility differences $\delta_{\mathcal{H}}^{m}$ of the human based on (observed) thresholds $\tau_{\mathcal{H}}^{m}$ in every stage m.

Note. *In the context of the automation continuously identifying the human density distribution while this identification influences the automation's strategy, Assumption 4.1 results from the well-known chicken-and-egg problem which is common for the identification of human cooperative behavior [Ing21, p. 2].*

The general procedure of the iterative identification algorithm for every stage m of the n-stage war of attrition is depicted in Figure 4.8 and explained in the following.

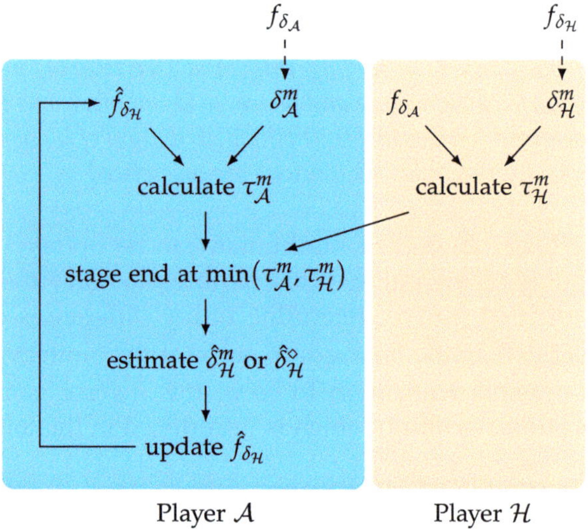

Figure 4.8: Overview of the iterative identification algorithm to solve the inverse game of the n-stage war of attrition.

At the end of stage m, i. e. one player has reached the individual threshold τ^{m} and gives in, the update of the density distribution $\hat{f}_{\delta_{\mathcal{H}}}$ depends on whether player $\mathcal{A}$ has won or lost the stage:

- if player $\mathcal{A}$ has won, she or he observes $\tau_{\mathcal{H}}^{m}$ and is able to estimate the utility difference $\hat{\delta}_{\mathcal{H}}^{m}$ based on (3.24a) or (3.24b), depending on the roles player $\mathcal{H}$ had at the beginning of stage m. Then player $\mathcal{A}$ updates the density distribution $\hat{f}_{\delta_{\mathcal{H}}}^{m}$ by

$$\hat{f}_{\delta_{\mathcal{H}}}^{m+1} \leftarrow \frac{m}{m+1} \cdot \hat{f}_{\delta_{\mathcal{H}}}^{m} + \frac{1}{m+1} \cdot \delta^{\dagger}(\hat{\delta}_{\mathcal{H}}^{m}) \tag{4.10}$$

  with Dirac delta function $\delta^{\dagger}(\cdot)$.

- if player $\mathcal{A}$ has lost and gave in at $\tau_{\mathcal{A}}^m$, she or he estimates the utility difference $\hat{\delta}_{\mathcal{H}}^{\diamond}$ that would have resulted in player $\mathcal{H}$ having the same threshold $\tau_{\mathcal{H}}^m = \tau_{\mathcal{A}}^m$. The estimation is based on (3.24a) or (3.24b), depending on the role player $\mathcal{H}$ had at the beginning of stage m. Then player $\mathcal{A}$ updates the density distribution $\hat{f}_{\delta_{\mathcal{H}}}^m$ with respect to $\delta_{\mathcal{H}}^m \in (\hat{\delta}_{\mathcal{H}}^{\diamond}, \infty)$:

$$\hat{f}_{\delta_{\mathcal{H}}}^{m+1} \leftarrow \frac{m}{m+1} \cdot \hat{f}_{\delta_{\mathcal{H}}}^m + \frac{1}{m+1} \cdot \frac{\hat{f}_{\delta_{\mathcal{H}}}^m \cdot \Theta(\hat{\delta}_{\mathcal{H}}^{\diamond})}{1 - \hat{F}_{\delta_{\mathcal{H}}}^m(\hat{\delta}_{\mathcal{H}}^{\diamond})} \tag{4.11}$$

  with $\Theta(\cdot)$ being the Heaviside step function.

Note. *The update rule weights all, past and current, observations equally due to the assumption that $f_{\delta_{\mathcal{H}}}$ is time-invariant.*

The first identification update rule (4.10) is motivated by the law of large numbers [BLP16], i. e. the expected value and the variance of the identification result converge for $m \to \infty$ to their ground truth values. However, this update is only applicable if a threshold of player $\mathcal{H}$ is observed. Therefore, the second update rule (4.11) tries to make use of the perceived information of $\delta_j^l > \delta_{\mathcal{A}}^{l:m}$ if player $\mathcal{A}$ gives in at stage m.

Remark. *The identification of the density distribution $\hat{f}_{\delta_j}$ at the end of every stage m may lead to situations in which the threshold $\tau_{\mathcal{A}}^{m+1}$ for the next stage $m+1$, calculated by means of the density distribution $\hat{f}_{\delta_j}^m$ updated at the end of stage m, becomes negative. This can be solved either practically by setting $\tau_{\mathcal{A}}^{m+1} = 0$ or by only updating the density distribution $\hat{f}_{\delta_j}$ at the end of a game taking into account all corresponding observations.*

Establishing the Cost Function

Another crucial design factor of the automation based on the n-stage war of attrition is the definition of the cost function, see Definition 3.13 and Assumption 3.12. The study on models' suitability (see Section 4.1) reveals that exponential functions with an average exponent of 1.95 fit human behavior. Therefore, $c(t) \sim t^2$ may be an appropriate initial choice in a practical application. However, the prefactor values of the exponential function are strongly influenced by the duration of the cooperative decision making process. Therefore, a general criteria for meaningful concession behavior of the automation is motivated by the threshold calculations (3.24a) and (3.24b): the potentially largest utility difference between the highest and lowest utilities within a decision scenario should be in the same order of magnitude as the cost functions value at the deadline i. e.

$$\max_{d_k \in D} u_{\mathcal{A}}(d_k) - \min_{d_k \in D} u_{\mathcal{A}}(d_k) \approx c(\mathcal{T}) \, .$$

This should yield a conceding behavior of the automation that is neither overly concessive nor dominant.

Note. *Sometimes the introduction of a soft deadline is favored to practically ensure a mutual agreement when reaching the deadline. To this end, a steep incline shortly before* $t = \mathcal{T}$ *may be added to the cost function.*

Remark. *In general without any consideration, there is no guarantee for an agreement until a given deadline. Therefore, the automation design also needs to address the allocation of ultimate authority to decide in non-agreement cases.*

In summary, the report on the suitability study and the above introduction of automation designs based on mathematical behavior models of cooperative decision making provide an answer to the second research question of this thesis, see Section 2.4. Furthermore, they also provide important insights and guidelines for a practical implementation of the automation designs. This includes the communication interface design which is essential for cooperative decision making and the design of decision scenarios for an experimental investigation of human-machine cooperation on decision level. On this basis, the following chapter presents two experimental evaluations of implemented automation designs based on the adaptive negotiation model and the n-stage war of attrition.

5 Experiments

This chapter reports on two experimental evaluations of the automation designs based on human-machine cooperative decision models which are proposed in the previous chapter. The experimental evaluation is a means to methodically compare the newly proposed automation designs with state-of-the-art approaches in practice. The results provide first evidence that the proposed emancipated human-machine cooperation on decision level outperforms state-of-the-art autonomy-centered and human-centered cooperation designs.

Both experiments consider a common and highly investigated application area: the human-machine cooperative control of highly automated mobile entities. The exemplary application scope of the first experiment is the teleoperation of mobile robots: the robot has two LOA, *manual control* and *automated control*, and the robot's automation and the human operator have to cooperatively and dynamically decide on the appropriate choice of LOA. The second experiment focuses on highly automated driving: human and machine have to cooperatively decide which driving maneuver to select which is then executed by the highly automated car.

As a prerequisite for these experiments, this chapter initially introduces a novel experimental evaluation approach focusing on the decision level of human-machine cooperation by discussing the corresponding challenges and measures.

All in all, this chapter provides an answer to the third research question of this thesis, see Section 2.4.

5.1 General Experimental Evaluation Approach for Human-Machine Cooperation on Decision Level

Although there is no experimental evaluation of human-machine cooperation exclusively focusing on the decision level, there are some experimental reports investigating human-machine cooperation which partially comprise decision making in some form [OKSB12, MLK$^+$12, DvA$^+$10, BAMF14, WWM$^+$19].

On this foundation, a general experimental evaluation approach for human-machine cooperation focusing on decision level is introduced in the following. It provides specific requirements for a suitable experimental design with respect to cooperative decision making and customized measures for an expressive experimental evaluation. Apart from these specific requirements and measures, the prevalent comparative character of experiments, i. e. comparing newly-proposed and state-of-the-art concepts, is also applied in the experimental evaluation of human-machine cooperation on decision level.

5.1.1 Measures for Experimental Evaluation and Comparison

In order to experimentally evaluate and compare automation designs for cooperative decision making, a set of measures considering both subjective user aspects and objective cooperative aspects is proposed. The measures are inspired by and aggregated from experiments conducted in the context of human-machine cooperation [OKSB12, MLK+12, DvA+10, BAMF14, WWM+19]. They are customized to suit the evaluation and comparison of cooperative decision making automation designs. All measures require a sufficiently large series of decision making scenarios between human and the respective automation design in order to yield meaningful results.

Subjective User Aspects

The following subjective aspects can be evaluated by means of a questionnaire which is typical for human-centered analysis of automation designs [OKSB12, MLK+12, BAMF14, WWM+19].

- **Satisfaction**
  How satisfied are humans with the cooperation in general?
- **Trust**
  How much do humans trust in the automation during the process of cooperative decision making?
- **Transparency/Reasonability**
  How subjectively transparent/reasonable do humans perceive the interaction with the automation?
- **Mental Load/Excitement**
  How mentally demanding do humans perceive the interaction with the automation?
- **Frustration**
  How frustrating do humans perceive the interaction with the automation?
- **Usability**
  How intuitive do humans perceive the interaction with the automation and corresponding interfaces?

Objective Cooperation Aspects

The following measures allow for an objective evaluation of the cooperation partners' performance and involvement in the cooperative decision making. However, all applied metrics have to be carefully designed according to the respective scenario of cooperative decision making in order to avoid in- and over-sensitivity towards any evaluated automation design.

- **Objective Cooperation Performance**
  Some objective metric that allows for measuring the performance of cooperation with respect to the given decision scenario which e. g. requires information fusion of both cooperation partners.
- **Balance of Conceding**
  The ratio between the numbers of instances each cooperation partner concedes.
- **Effort**
  A metric that evaluates the effort of the cooperation partners in a given cooperative decision making scenario, e. g. in terms of communication.

Partially based on the proposed measures, the following section composes requirements on the experimental design for evaluating human-machine cooperative decision making.

5.1.2 Requirements on the Experimental Design

In general, experiments investigating human-machine cooperative decision making have to feature decision scenarios which are plausible and intuitive for humans and allow for a suitable application of all or a subset of the evaluation measures introduced above [RWIH20]. To this end and with respect to the meta-model of human-machine cooperative decision making introduced in Section 3.1.4, the following list provides more detailed requirements on a suitable experimental design for human-machine cooperative decision making. This list of requirements is referenced in all following explanations of experimental designs to ensure consistent experimental designs.

a) A controlled but realistic and dynamic environment with a sufficiently large amount of decision scenarios has to be provided.

b) The decision scenarios have to be intuitively comprehensible by participants, e. g. the number of decision options should be small, see Section 3.1.2.

c) Each decision scenario has to comprise a set of decision options and allow (sufficiently often) for differences in decision option preferences between human and automation which lead to decision conflicts. Furthermore, for reasons of practicality, a pressure for making a (consensual) cooperative decision in conflict situations should be present.

d) A human-machine cooperation on decision level has to be enabled, i. e. no cooperation partner is always able to outperform the other in terms of reasonable decision making. Furthermore, cooperation partners have to be able to communicate during cooperative decision making.

e) Suitable objective and subjective measures to evaluate cooperative performance as introduced in Section 5.1.1 have to be defined.

f) Confounding variables especially considering human-machine communication have to be avoided.

g) The repeatability of experimental runs considering the different investigated newly-proposed and state-of-the-art concepts has to be ensured.

h) Learning effects during the course of the experiment have to be avoided.

Taking these requirements and proposed measures of the general evaluation approach for human-machine cooperative decision making into account, the following sections report on two experiments conducted to evaluate the automation designs based on the models of human-machine cooperative decision making proposed in Chapter 3.

5.2 Cooperative Decision Making in Mixed-Initiative Control of Robots

The following experimental report on cooperative decision making in mixed-initiative control of robots is the result of a collaboration with the *Extreme Robotics Lab* at the University of Birmingham (United Kingdom) and is currently in the publishing process [RCI$^+$22].

In recent years, the control of mobile robots has evolved from sole manual teleoperation to assisted teleoperation to robots with a variable LOA. For assisted teleoperation, concepts such as shared control have been applied for manipulation tasks, e. g. [CSP14, MLH15]. In essence, these approaches use some form of input mixing or policy blending between the robot's controller and/or the operator's control inputs [DS13]. Control conflicts arise when the desired trajectories of the operator differ from the automation controller's assistive trajectories, e. g. the controller induces guiding forces contrary to the human's desired movement [MO04]. To tackle this problem, researchers utilize trajectory learning and intention recognition strategies [KSB13, JWBA16]. Hence, these assistive teleoperation systems adapt their level of assistance.

In contrast, robots with a variable LOA may cause a different form of conflict for control, i. e. the human operator and the robot prefer different LOA. Although there are some approaches which avoid these conflicts by recommending or asking the operator for LOA switching or other actions [TI17, HIL$^+$19, CRDD20], these variant LOA systems usually allow switches between different LOA: both the operator and the robot's automation have the authority to initiate or completely override each other's commands in a variety of levels of abstraction, e. g. from direct control commands to role and task assignment [CHS21]. The basis of these systems is denoted as mixed-initiative (MI) control which is defined as "a collaboration strategy for human-robot teams where humans and robots opportunistically seize (relinquish) initiative from (to) each other as a mission is being executed, where initiative is an element of the mission that can range from low-level motion control of the robot to high-level specification of mission goals [. . .]" [JA15]. In this thesis, MI control refers to the authority of both the robot's automation and the operator to initiate LOA switches.

Existing work often tackled potential conflicts for control rather reactively and intrusively by the robot's automation taking control triggered by specific (usually safety-critical) events [NFA08, HG09, VGLH11]. Only a few approaches tried to properly resolve the conflict for control. Mercier et al. [MTD10] proposed an authority dynamics controller based on a dependence graph of resources, such as the robot's wheels or its pose. These resources could be controlled by either the operator or the robot. They solved authority conflicts by reallocating these resources based on task-specific predefined authority priorities. Owan et al. [OGD17] proposed a consensus procedure based on heuristically determined timeout thresholds to solve control conflicts. When consent could not be reached, similarly to [MTD10], a task-specific heuristic contingency procedure was triggered based on predefined authority priorities.

In summary, variable LOA systems (including MI systems) found in literature often do not use any explicit policies for avoiding conflicts. They either ask for the operator's help when an autonomy level modification is needed (e. g. the operator taking control) or intrusively take the initiative. The few works offering explicit policies for dealing with authority transfer and conflicts are based on predefined priorities which agent has authority in which scenario.

Therefore, the following sections report on an experiment comparing the state-of-the-art *expert-guided mixed-initiative control switcher* (EMICS, introduced in [CHS21]) with the newly proposed *negotiation-enabled mixed-initiative control switcher* (NEMICS). The NEMICS is a novel MI control system which is enabled to cooperatively and explicitly resolve conflicts for control by means of utilizing the basic negotiation model (see Section 3.2.3) from the adaptive negotiation model of Section 3.2. This was the first step of experimentally investigating research on human-machine cooperative decision making. Additionally, this was the first effort of the research collaboration to gain some initial experience of introducing negotiation theory to MI control switcher design. The cooperative performance of human operators with the NEMICS was evaluated and compared to the cooperative performance with the EMICS by means of subjective and objective cooperative performance measures: operators' frustration, time to reach the destination, number of collisions, and number of conflicts. The corresponding hypothesis was the following.

> **Hypothesis 5.1 (Subjective Assessment and Objective Performance)**
> *The application of NEMICS in comparison to EMICS leads to a reduced operators' frustration and increased objective cooperative performance in terms of smaller times to destination and reduced number of collisions and conflicts.*

Subsequent to the introduction of the experimental design in Section 5.2.1, Sections 5.2.2 and 5.2.3 report the experiment's results and discuss these findings.

5.2.1 Experimental Design

The experiment was designed along the requirements on experimental designs in the context of human-machine cooperative decision making introduced in Section 5.1.2:

a) The basis of this experiment was the simulation of a mobile robot operating in a realistic *search-and-rescue scenario* [CTS19].

b) For navigating the robot towards a predefined destination, there were two decision options (i. e. LOA): either the robot navigated autonomously or the robot was navigated via teleoperation.

c) While navigating through the simulated environment, changing circumstances incentivized the robot's MI control switcher and the human operator, i. e. the two cooperation partners, to continuously decide whether or not a LOA switch would be appropriate. The incentives for switching the LOA resulted from temporarily different navigational objectives and performances of the robot's automation and the human operator: the temporarily navigational objective was the intuitive investigation of human victims along the path to the destination. Furthermore, the robot's automation and the human operator faced realistic navigational performance degradation in the form of sensor noise and secondary tasks, respectively.

d) As a result, conflicts for control arose which had to be solved and no cooperation partner was able to always outperform the other in reasonable decision making.

e) The cooperative decision making was evaluated by means of a subjective usability questionnaire and the NASA-TLX [Har06] assessing the participants' frustration. Furthermore, the time to reach the destination, the number of conflicts for control, and the number of collisions were utilized as objective cooperative performance measures.

f) The communication between participants and the MI control switchers was based on an intuitive two-button, graphical and acoustical interface.

g) Participants had to navigate twice through the search-and-rescue scenario while facing one of the two different MI control switchers, either the state-of-the-art EMICS or the newly proposed NEMICS, in separate experimental runs.

h) To counteract potential learning effects, the sequence in which EMICS and NEMICS were active was counterbalanced among participants and participants got to know the setup and simulation environment in a standardized training in advance of the actual experimental runs.

The following sections introduce experimental setup, the conflict for control scenarios, the experimental procedure, the applied measures as well as the applied MI control automation designs EMICS and NEMICS in more detail.

Setup

The experimental setup consisted of a mobile robot simulation and an *operator control unit* (OCU) which allows for the interaction of a human operator with the simulated robot in a search-and-rescue scenario.

The environment and the robotic system were simulated in *Gazebo*, a high fidelity robotic simulator. The simulated robot was a mobile robot, the *Clearpath Robotics Husky Unmanned Ground Vehicle*, equipped with a laser range finder and a camera. It was capable of operating in two different types of LOA: teleoperation (operator fully in control of navigation via the OCU) and autonomy (autonomous navigation towards a predefined destination). The software of the MI control framework and related capabilities was developed by means of the *robot operating system* (ROS) and is described in detail in [CSB$^+$16, CHS21].

A simulated environment was chosen to avoid introducing complex confounding factors from a real robot operating in the real world and for improving the experiment's repeatability. As it can be seen in Figure 5.1, the simulation environment created very realistic situations and stimuli for the participants as experienced when operating a real robot. In addition to the experiment's test environment, a similar training environment was provided for the participants to become familiar with the hardware setup and simulated robot. Both environments were approximately 720 m^2 of similar difficulty but different layout.

The robot was controlled via the OCU which was composed of a joypad as an input device, a laptop running the software of the MI control framework and for simulating the environment, and a screen showing the *graphical user interface* (GUI), see Figure 5.2. To navigate the robot in teleoperation mode, the direction controller on the joypad was used. Additionally, the operators could communicate their choice of LOA via two buttons on the joypad: if interacting with EMICS, this led to a LOA

Figure 5.1: The simulation environment of the search-and-rescue scenario used for the experimental evaluation.

Figure 5.2: The graphical user interface. **Left:** video feed from the robot's camera (1), the control mode in use (2) and the status of the robot (3). **Right:** The map (4) showing the position of the robot, the current destination (blue arrow), the optimally planned path (green line), the obstacles' laser reflections (red) and the walls (black). **Bottom:** The negotiation display (disabled if only EMICS is active) with the available control modes (left: autonomy (5), right: teleoperation (6)) and a bar graph (7) to visualize the remaining negotiation time.

switch; if interacting with NEMICS, this either initiated or was part of a negotiation whether or not to switch the LOA.

The negotiation part of the GUI consisted of an image of the teleoperation LOA and the autonomy LOA and a bar graph to visualize the negotiation deadline, i. e. the remaining time for negotiation. All elements of the negotiation display were in standby mode (black overlay) if agents were not negotiating. If a negotiation was active, LOA choices of EMICS and human operator were visualized by different background colors of the respective LOA images (blue - choice of EMICS; orange - choice of operator) and the remaining negotiation time was depicted by the red portion of the bar graph. If an agreement was reached, the agreed LOA was highlighted with green color while all other elements returned to the standby mode. After 3 s, all elements were in standby mode again. This negotiation GUI had been successfully applied in the suitability study reported on in Section 4.1 and in [RWIH20].

Conflict for Control Scenarios

The experimental scenario was composed of six areas depicted in Figure 5.3. The primary task objective of the human-robot system was to navigate from Area 1 to the destination in Area 6 as quickly as possible while avoiding collisions. The remaining four areas were designed to evaluate the functionality of EMICS and NEMICS in various LOA switching situations with potential conflicts for control. These situations were created by introducing secondary objectives or performance degrading factors.

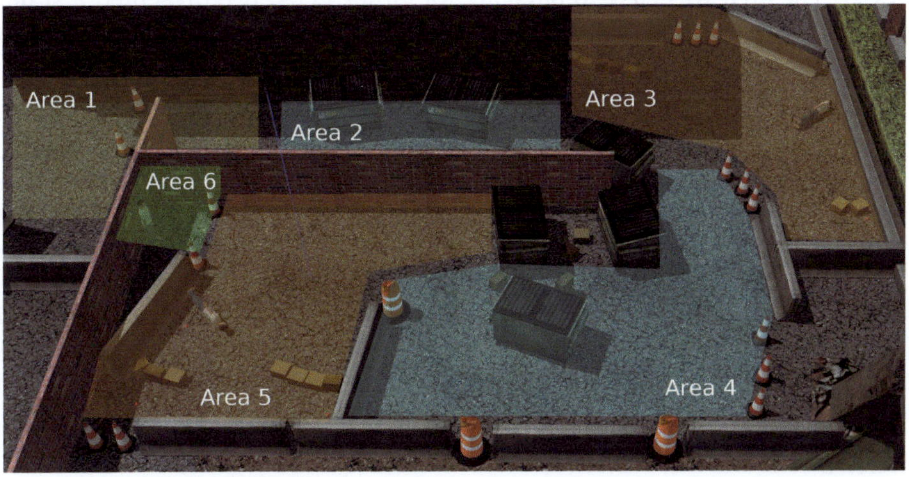

Figure 5.3: The conflict for control Areas 1 to 6 in the simulated environment of the search-and-rescue scenario. ©2022 IEEE

While participants were performing the primary task, their secondary objective was to spot a human victim in both Areas 3 and 5. Each of these human victims was associated with three *points-of-interest* (POI) represented by three red balls that participants had to locate. The location of the balls had been unknown to the participants in advance. Each POI was considered completed when the ball was entirely covered by the laser's mounting, visible in the lower center of the camera's video feed. This incentivized equal proximity of each participant to each ball. Localizing the POI caused a detour and ultimately led to conflicts for control with the MI control switcher due to opposing objectives. While locating the POI, some obstacles were undetectable by the robot but visible to the operator via the camera feed. Hence, they were additional source of potential conflicts for control concerned with avoiding collisions. While navigating through Area 2 and 4, the human-robot system experienced situations of performance degradation of either the robot's automation through artificial sensor noise or the operator through a math task of adding a series of 3-digit numbers. The sensor noise and the math task began when the area was entered and were lasting for 15 s each. During the period of performance degradation of one agent, the other agent had an incentive to take control. In this case, it was assumed that the agent with degraded performance would not oppose the other agent taking control and hence no conflict for control was expected.

The following listing provides more details on the six areas constituting one experimental run.

- **Area 1**
  This was the starting area with the robot initially operating in the autonomy LOA. The area was easy to navigate for either LOA. It represented a situation without any incentive for the MI control switcher or the operator to initiate a LOA switch.

- **Area 2**
  As the robot entered this area, artificial noise was introduced to the laser scanner readings to degrade autonomy's performance. As a result, if autonomy LOA was active, the robot's autonomous navigation was slowing down. However, the noise was not enough to make the MI control switcher initiate a LOA switch. It was expected that the operator would like to overcome the performance degradation and hence would initiate a LOA switch to teleoperation. Consequently, this area represented a situation in which the operator had an incentive to initiate a LOA switch while the MI control switcher had no incentive to resist.

- **Area 3**
  This area was easy to navigate for either LOA. The operator could spot a human victim and was asked to inspect it and its close-by POI, i. e. the red balls. Hence, if the autonomy mode was active, the operator had an incentive to

change to teleoperation, which the MI control switcher would initially not oppose. Furthermore, the robot would then deviate from the expected path and the MI control switcher, inferring that the performance has dropped, would initiate a LOA switch to autonomy. This led to a situation where the operator had an incentive to persist on her or his chosen LOA (exploring the POI with teleoperation) while the MI control switcher insisted on an opposing LOA (reducing the path deviation via giving control to autonomy). This is the kind of situation in which typically conflicts for control emerge as observed in [CHS21]. After the inspection of all red balls, the operator was expected to return to the original path.

- **Area 4**
  Within this area, the human operator was asked to conduct the math task, hence the operator's performance (or capacity for performing well), was expected to decrease. As a result, if the teleoperation was active, either the operator or MI control switcher would initiate a LOA switch to autonomy. This represented a situation in which the operator and the MI control switcher had an incentive to switch to the same LOA.

- **Area 5**
  This area is similar to Area 3 being easy to navigate for either LOA. The operator could spot a human victim and was asked to inspect it and its close-by POI. Hence, if the autonomy mode was active, the operator was expected to initiate a LOA switch to teleoperation. The MI control switcher had no incentive to oppose strongly. As a result, the teleoperated robot would deviate from the expected path while the MI control switcher inferred the operator's performance degradation and initiated a LOA switch to autonomy. This again led to a situation where the operator had an incentive to persist on her or his chosen LOA while the MI control switcher insisted on an opposing LOA. After the inspection of all POI, the operator was expected to return to the original path.

- **Area 6**
  This was the destination area in which the experimental run was terminated.

Note that the operator and EMICS were able to freely initiate LOA switches at any moment. In the case of using NEMICS, the operator and EMICS were able to freely initiate negotiations for LOA switches.

In summary, there were two areas with an expected conflict for control due to different objectives of the operator and MI control switcher and three non-conflict situations in which both agents did not have an incentive to oppose the other's wish for switching LOA.

Mixed-Initiative Automation Designs

In the following, the state-of-the-art EMICS and the novel NEMICS are introduced.

The EMICS uses an expert-guided approach to initiate LOA switches [CHS21]. It assumes the existence of a task expert (e. g. a navigation planner) which, given a navigational destination, is able to provide the expected task performance for the human-robot system in the absence of performance-degrading factors. The comparison between the system's run-time performance with the expected expert performance yields an online task effectiveness metric called goal-directed motion error[21] $g \in [0,1]$ [CHS21]. In essence, the error describes the difference between the robot's current motion and the motion of the robot required to reach the destination according to the expert planner. Hence, the error metric expresses how effectively the system performs the navigation task. On this basis, the EMICS infers whether a LOA switch is beneficial. In practice, the EMICS's error thresholds were trained by observing human operators in previous experiments. The EMICS informs the operator about the initiated LOA switch using an alarm sound identical to the one denoting autopilot disconnection in aircraft, a synthetic speech expressing the LOA the system switched to, and a GUI notification.

Two assumptions are key in the design of EMICS: the human operator is willing to be in control and to hand over control based on the initiative of the EMICS, and the agent to which the control will be handed (i. e. either the human or the MI control system) is capable of correcting the task effectiveness degradation as expressed by the error. These assumptions have been found to cause conflicts for control in situations where the operator has different navigational objectives or information than the EMICS. In such cases, the EMICS infers a performance drop due to an increased error. At the same time, operators try to follow their navigational objectives or information which are unknown to the robot. As EMICS and operator have the same authority to switch LOA, this results in a series of conflicts for control, i. e. aggressively overriding the other's LOA switches.

In contrast to this, the novel NEMICS enhances state-of-art MI control, e. g. EMICS, by adding negotiation capabilities to address conflicts for control. By means of this approach, any MI control switcher can be enhanced as long as it provides some sort of utility measure for the different decision options (in this context LOA). The resulting framework enables the robot's automation and the human operator to negotiate the LOA during operation by means of a negotiation interface, i. e. the negotiation module, that allows for the communication and negotiation of the desired LOA.

The relation of robot, NEMICS and operator is depicted in Figure 5.4, also illustrating the advancement of the EMICS by means of the negotiation module towards NEMICS. The proposed negotiation module in NEMICS was designed according to a basic negotiation model introduced in Section 3.2.3: Two agents, i. e. the NEMICS

[21] Referred to as *error* for the rest of this section.

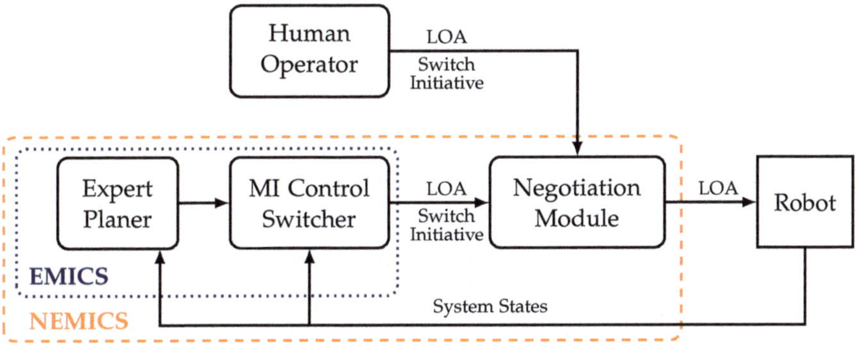

Figure 5.4: Block diagram of EMICS and NEMICS and their interaction with the robot and human operator. ©2022 IEEE

($\mathcal{A}$) and the human operator ($\mathcal{H}$), exchange offers which resemble the decision option, i. e. the different types of LOA which are teleoperation and autonomy. This set of offers, i. e. decision options, $O = \{\text{autonomy, teleoperation}\}$ is selectable via the interface. Both agents are able to freely initiate a LOA negotiation if they want to switch the LOA by proposing the other LOA via the interface. While negotiating, the agents are allowed to propose offers, i. e. concede to the other LOA offer, at any time, see asynchronous negotiation protocol in Section 3.2.2.

The normalized utility function $\bar{u}_{\mathcal{A}} \in [0, 1]$ enables NEMICS to evaluate the current LOA $o \in O$ by means of the normalized error metric $g \in [0, 1]$ of the EMICS, see explanations on the error metric above and in [CHS21]:

$$\bar{u}_{\mathcal{A}}(o) := \begin{cases} 1 - g & o \text{ represents active type of LOA} \\ 0.8 & o \text{ represents inactive type of LOA} \end{cases} \tag{5.1}$$

Note that the utility estimation of the inactive type of LOA is a difficult, predictive task. Since this was not the focus of this experiment this problem had been simplified: assuming a constant utility value for the inactive type of LOA reflects both the hesitation to change LOA and the hope for improvement by means of a LOA switch.

The human-like concession strategy $\mathcal{E}_{\mathcal{A}}$ is time-based, see Section 4.1 and [RWIH20]. In starting or joining a negotiation, NEMICS always starts to offer the LOA with the highest normalized utility $o^0_{\mathcal{A}} = \arg\max_{o \in O} \bar{u}_{\mathcal{A}}(o)$. In case of a conflict, it was assumed that there was a negotiation deadline $\mathcal{T}$ in place for practical reasons until which NEMICS and the human operator were required to agree on one LOA. Therefore, NEMICS concedes towards the other LOA if a decreasing, normalized target utility $\bar{u}_{t,\mathcal{A}}(t)$ has diminished by more than the normalized utility difference between the two LOA utilities $\Delta\bar{u}_{\mathcal{A}} = \max_{o \in O} \bar{u}_{\mathcal{A}}(o) - \min_{o \in O} \bar{u}_{\mathcal{A}}(o)$. To this end,

NEMICS continuously evaluates the following condition:

$$1 - \Delta\bar{u}_{\mathcal{A}} > \bar{u}_{\mathrm{t},\mathcal{A}}(t) := 1 - \frac{t}{\mathcal{T}}^{1/\epsilon} \tag{5.2}$$

with $t \in [0, \mathcal{T}]$ and the concession parameter ϵ. NEMICS concedes if this condition does no longer apply.

With two decision options available, the maximum negotiation time was set to $\mathcal{T} = 4\,\mathrm{s}$ which was enforced by the provided interface and motivated by 2 s reaction time per decision option. This deviation from the recommended 3 s reaction time per decision option (see Section 4.2.1) was motivated by the low number of decision options and the easy to use decision making input device.

Procedure

Each participant was introduced in a standardized manner (see [CTS19]) to the hardware setup and the simulation environment by operating the robot in a training environment for ten minutes. Hence, participants became familiarized with the robot's driving behavior, performance degradation, the LOA switching behavior when interacting with EMICS or NEMICS and the ball-locating task in the context of the POI exploration.

After the training, participants were informed about the upcoming two experimental runs and about their general objectives. For the two experimental runs, EMICS and NEMICS were employed separately. The sequence order of the EMICS and the NEMICS was counterbalanced among participants to compensate the influence of learning effects. Additionally, the layout of POI was such that operators were restricted from using different exploration strategies or paths and hence restricting individual variability. After conducting the two experimental runs, participants were asked to file the NASA-Task Load Index (TLX) questionnaire [Har06] once for each experimental run and a usability questionnaire to compare NEMICS and EMICS.

Measures

To evaluate the performance of the newly introduced NEMICS and compare it with the EMICS, the following objective measures were considered:

1) the time-to-completion of the primary task,
2) the number of collisions with the environment as a measure of safety, and
3) the number of conflicts for control in EMICS and the number of negotiations in NEMICS as a measure of human-robot-interaction performance.

The conflict for control is defined as a situation in which the EMICS and/or the operator aggressively override each other's LOA choices. For example, a situation in which the operator is in teleoperation LOA and the EMICS switches to autonomy LOA, forcing the operator to switch back to teleoperation, counts as one conflict. Similarly, a successful negotiation is defined as a situation in which the NEMICS has successfully negotiated an LOA switch that would otherwise result in a conflict.

Additionally, the NASA-TLX questionnaire [Har06] was applied as a subjective measure of the perceived workload level of operators when interacting with EMICS and NEMICS. Furthermore, a free form qualitative usability questionnaire was utilized considering user acceptance, intuitiveness, and transparency of interaction. The specific questions were:

Q1: *Was the interaction with either system intuitive?*

Q2: *Was the LOA switching behavior of either system transparent?*

Q3: *Was the LOA switching of either system intrusive?*

Q4: *Is there anything that could improve the LOA switching capabilities of either system that you can think of?*

Q5: *Anything that you would like to comment or add?*

Participants

A total of 10 participants took part in the study, 9 males and 1 female with a mean age of 31.5 years. All of them were experienced robot operators with extensive previous experience operating similar robotic systems.

5.2.2 Results

Given the relatively small sample size, the following presentation of the experiment's results focuses on the descriptive statistics and the qualitative results. The descriptive statistics for the objective measures and the NASA-TLX score can be found in Table 5.1.

There is a trend of participants completing the navigation task faster when using the NEMICS ($M = 231.4\,\text{s}, SD = 16.2$) compared to the EMICS ($M = 238.4\,\text{s}, SD = 23$). Participants had more collisions when using the EMICS ($M = 1.8, SD = 1.7$) compared to NEMICS ($M = 0.8, SD = 1.2$). While using the EMICS 12 out of the in total 18 collisions took place during conflicts. While using the NEMICS 1 out of the 8 collisions took place during the negotiations. Furthermore, a higher number of conflicts for control with EMICS ($M = 8.7, SD = 2.3$) was observed than numbers of successful negotiations with NEMICS ($M = 7.1, SD = 1.6$) that avoided potential

Table 5.1: Objective measures' results time-to-completion, number of collisions and number of conflicts for control (EMICS) or of negotiations (NEMICS), and NASA-TLX scores.

Participant	Time in s		No. of collisions		No. of conflicts		NASA-TLX	
	NEMICS	EMICS	NEMICS	EMICS	NEMICS	EMICS	NEMICS	EMICS
1	228	237	1	3	6	7	42.5	61.7
2	215	207	0	0	6	7	34.2	55.8
3	218	233	0	1	10	12	38.3	61.7
4	227	257	3	4	6	12	35.0	81.7
5	250	255	0	1	6	7	40.8	42.5
6	221	215	0	0	8	6	30.8	47.5
7	238	263	3	5	7	9	26.7	58.3
8	213	213	0	1	5	7	46.7	19.1
9	262	274	1	1	9	11	56.7	50.0
10	242	230	0	2	8	9	35.8	38.3
M	231.4	238.4	0.8	1.8	7.1	8.7	38.8	51.7
SD	16.2	23.0	1.2	1.7	1.6	2.3	8.5	16.7

conflicts for control. Participants experienced a higher cognitive workload leading to higher NASA-TLX scores while using the EMICS ($M = 51.7, SD = 16.7$) compared to using the NEMICS ($M = 38.8, SD = 8.5$).

Regarding the usability, 9 out of 10 participants found the interaction with both systems (i. e. EMICS and NEMICS) intuitive, see Q1. However, 5 out of these 10 participants stated that the NEMICS was more intuitive than EMICS, 4 participants found EMICS more intuitive, and one participant perceived both systems equally intuitive.

Considering Q2, 3 out of 10 participants found the LOA switching behavior of both systems to be equally transparent, 6 out of 10 participants perceived the NEMICS to be more transparent, and only 1 participant found EMICS to be transparent, but not NEMICS.

Considering Q3, 8 out of 10 participants found EMICS to be more intrusive compared to the NEMICS. One participant perceived the NEMICS more intrusive than the EMICS and one participant found both MI control switchers to be equally intrusive.

Regarding the objective performance results and the subjective assessment of the participants, evidence was found which supports Hypothesis 5.1.

Furthermore, the usability questions (see Q4 & Q5) have provided important insights. First, participants thought that the negotiation method and respective way of communication with the operator was an improvement compared to the more intrusive hand-off strategy of the EMICS, e. g. "NEMICS was much less intrusive but still, some interaction was needed, having a grace period [meaning to negotiate] helped",

"NEMICS was more intuitive as you expect from a robot to negotiate and listen to you", and "NEMICS was an improvement over EMICS." However, participants also stated "because of tunnel vision and concentration on the task you might miss a negotiation" and "negotiation is still intuitive but the GUI is complex, [provides] too much info".

Second, participants expressed the view that they should have a more direct and instant influence on negotiation e. g. "[I would like] instant negotiation in some cases, e. g. when operator wants control [but not when the robot wants control]." (at least 4 participants made similar statements).

Third, participants expressed the need to be better understood by the robot's automation to minimize the frequency of negotiations, e. g. "I want the robot to better understand what I want, understand that I was looking for the balls and not have to communicate the LOA multiple times" and "[I would like the robot to] understand intentions or tell the robot what you are doing."

5.2.3 Discussion

The trend to higher time-to-completion with the EMICS further strengthens the idea that this may be due to the conflicts for control as also observed in [CHS21]. Based on the observations, two factors negatively influence time-to-completion: the extra commands needed (i. e. extra LOA switches and extra maneuvers to correct for movement during the conflicts); and the higher cognitive workload as measured by NASA-TLX.

The mixed results considering the intuitiveness and transparency of the interaction (see Q1 and Q2) might be explained by the participants not being sufficiently aware of the start of a negotiation. As one participant suggested, one could "have a beeping sound once the negotiation started that stops once you made your LOA choice" to improve NEMICS.

Evidence from the study suggest that intrusive control authority transfer can lead to decreased safety in navigation as most of the collisions observed while using the EMICS were due to the conflict for control. While the operators were fighting for control with the EMICS, they could not concentrate on obstacle avoidance which is especially severe due to the (to them potentially undesired) maneuvering which happened in autonomy mode. Avoiding collisions was even more difficult as some boxes during the search task were not visible by the robot's sensors, and hence autonomy LOA would not avoid them. Additionally, the majority of participants also subjectively perceived EMICS as more intrusive, see Q3.

To further increase the usability of the NEMICS, the application of the entire adaptive negotiation model is expected to improve performance as it offers the capability to adapt to the human operators' negotiation behaviors, i. e. operators' actions during the negotiation. Furthermore, the evidence suggests that human intent recognition can play a crucial role in human-robot teaming and MI systems, potentially increasing user acceptance drastically.

Lastly, this experiment demonstrated the ability of NEMICS to deal with conflict for control due to unforeseen circumstances such as performance degrading factors for both agents and a mismatch in their objectives. Due to the realistic experimental design, the observed results motivate future investigations with real robots.

5.2.4 Conclusion

An experimental study was conducted, inspired by a search-and-rescue scenario in which a human-robot system had to navigate and search for points of interest. The mobile robot was controlled by a robot's automation and a remote human operator in a mixed-initiative manner. In the course of the experiment two MI control strategies were compared: the state-of-the-art EMICS with the newly proposed NEMICS based on negotiation theory, see Section 3.2.3.

This study provides the first experimental evidence that the application of a negotiation model enabling the robot to cooperatively make a decision on the appropriate LOA reduces conflicts for control and can potentially counteract their negative effects on cognitive workload, operational performance and safety metrics. Furthermore, the study's results highlight again how crucial an adequate interface and decision scenario design is to enable intuitive cooperative decision making.

The success of NEMICS encourages future investigations of applying the entire adaptive negotiation model and the n-stage war of attrition in similar MI control switcher designs. Furthermore, this success is assumed to be generalizable to other scopes and realistic implementations due to the general and realistic experimental setup. Therefore, the next section examines both automation designs based on the adaptive negotiation model and on the n-stage war of attrition in the application scenario of highly automated vehicles.

5.3 Cooperative Decision Making in Highly Automated Driving

The experiment reported on in this section is currently under review for publication [RWI^{+}] and was conducted in the course of a master thesis [Wör20]. The experiment focuses on cooperative decision making in the application scenario of a highly automated vehicle. Resulting from an increasing degree of automation in vehicle control, guidance and navigation in form of already available advanced driving assistance systems, the driver's role changes continually from manual (assisted) control towards supervision of the automated driving systems [FKGH15, FCA+17, ACM+18, Fla19, WLCW19]. Research has revealed that drivers become increasingly unaware of the driving situation if no (supervisory) action of them is required [GDLB13, FBB+14, End17]. Hence, engineers of driving assistance systems face the general "out-of-the-loop performance problem" [EK99] which can be observed for users interacting with any highly automated systems: in case human action is required at some point due to e. g. functionality boundaries of the automated system, the human, in this case the driver, is almost certainly unable to act appropriately due to lacking situation awareness. One approach is to carefully design the transition from automated driving back to manual driving by means of gradually shifting the control authority in accordance to the driver awareness [LHFH18]. Another approach is to keep the human in the loop at a higher task level, i. e. instead of conventional manual vehicle control, the driver operates the system on e. g. the guidance level (see Section 2.2.5, [FBB+14]), i. e. by means of maneuver commands [FKGH15, WLCW19]. In this context of keeping the driver in the loop while operating a highly automated vehicle, this experiment investigated emancipated human-machine cooperative decision making concerned with the maneuver selection. Although some research and approaches exist which consider dynamic authority assignment and/or offer decision support, the state of the art in cooperative decision making in this application context is the leader-follower approach with the human in the lead in non-critical situations, see Section 2.3.2. Therefore, this experiment compares the two automation designs based on the newly proposed cooperative decision making models (the adaptive negotiation model and the n-stage war of attrition, see Sections 3.2, 3.3, and 4.2) with the two leader-follower-based automation designs (human in lead while the automation follows, and vice versa). The comparison's evaluation was conducted with respect to objective measures and subjective assessment and investigated the following hypotheses.

Hypothesis 5.2 (Objective Performance)

The objective performance of the human-machine cooperation on decision level with automation designs based on cooperative decision making models is significantly better compared to the state-of-the-art leader-follower-based automation designs.

Hypothesis 5.3 (Subjective Assessment)

The participants' subjective assessments are significantly better for the proposed automation designs based on cooperative decision making models than for the state-of-the-art leader-follower-based automation designs in terms of satisfaction and trust in the cooperation as well as intuition of interaction. The opposite is expected regarding the transparency of interaction.

The following report on the experiment is structured as follows: The experimental design and evaluation of the automation designs' comparison is provided in Section 5.3.1 and 5.3.2, respectively. This is followed by the discussion of the results in Section 5.3.3 and some concluding remarks in Section 5.3.4.

5.3.1 Experimental Design

The experiment was designed according to the requirements on experimental designs in the context of human-machine cooperative decision making introduced in Section 5.1.2:

a) The experiment was set in a futuristic yet reasonable highly automated driving scenario (cf. similar research on "conduct-by-wire" [FKGH15]): A driving simulator depicted in Figure 5.5 was utilized to realistically recreate a drive in a highly automated vehicle through a so called *Manhattan grid*.

Figure 5.5: Front view of the driving simulator for highly automated driving equipped with three vehicle visualization screens (top), steering wheel and pedals (middle, unused in this experiment), a driver's seat (bottom) and a touchscreen as a maneuver decision interface (mid-right). ©2022 IEEE

b) The Manhattan grid comprised multiple intersections, each representing a cooperative decision scenario in which a driving direction (*left*, *right*, *straight ahead*) had to be chosen.

c) This choice was influenced by potential traffic delays associated with specific vehicles at different directions and the objective to minimizing travel time to reach a defined destination displayed on a map.

   Additionally, participants were made aware of their cooperative decision making performance with the automation by means of an objective performance measure based on the minimal travel time: after each decision made at an intersection the deviation in travel time between the chosen direction and the optimal choice was displayed as well as the overall time deviation between the optimal and chosen path.

d) In order to create decision conflicts, the human participant was only aware of the traffic at the upcoming intersection (local information) whereas the automation possessed information on all traffic (global information). However, the information of the automation was partially false (e. g. due to inaccurate perception of the dynamic traffic development at the upcoming intersection) such that the cooperation of human and automation, i. e. the intervention of the human, potentially yielded benefits which were observable via the displayed objective performance measure.

e) The cooperative performance was objectively evaluated by means of the travel time and subjectively via a questionnaire assessing the satisfaction with the cooperation, the intuition of interaction, and the reliability and transparency of the partner's behavior.

f) To avoid confounding factors, the human and the automation solely communicated via a cooperative maneuver decision making interface (CMDI) consisting of a head-up display and a touchscreen. The interface provided the discrete decision options of the next intersection, i. e. the available driving maneuvers, and the remaining time until a final decision had to be reached before entering the intersection. When reaching the intersection the (potentially) mutually chosen decision option, i. e. the driving maneuver, was executed by the highly automated vehicle which then continued the autonomous drive until the next decision scenario took place at the next intersection.

g) For each drive through the Manhattan grid, i. e. each experimental run associated with the application of a different decision making automation design, participants were confronted with the same Manhattan grid setup but were unaware as the displayed map was rotated by a defined multiple of 90°. Furthermore, the sequence of experimental runs within the experiment was randomized for each participant. In addition, participants got to know the experimental setup by means of a training phase in advance of the actual experimental runs.

In total, there were four experimental runs investigating the benefits of human-machine cooperation on decision level by comparing the two automation designs based on the cooperative decision making models proposed in Chapter 3 with the two manifestations of state-of-the-art leader-follower automation designs, i. e. either the human or the automation is in the lead. To easily differentiate the four automation designs in the following, the following abbreviations apply:

NT: automation design based on **N**egotiation **T**heory, i. e. on the adaptive negotiation model, see Section 3.2,

GT: automation design based on **G**ame **T**heory, i. e. on the n-stage war of attrition game model, see Section 3.3,

LH: automation design based on the leader-follower approach with the **L**eader being the **H**uman or

LA: the **L**eader being the **A**utomation.

Setup

The experiment's setup was based on a simulator for highly automated driving developed by the Institute of Control Systems (IRS) at the KIT with a human-machine interface on driving maneuver level for cooperative decision making, see Figure 5.5. Its core was a *XPACK4* real-time system from *IPG Automotive GmbH* and their vehicle simulation software *CarMaker® 8*. This setup was utilized to simulate the driving behavior of a car and its environment including traffic. For this experiment the hardware setup was enhanced by three visualization screens displaying the simulated vehicle, its surroundings and a head-up display as the visual part of the CMDI. Furthermore, a touchscreen was integrated on the right hand side of the driver's seat as active part of the CMDI. Additionally, a sound system provided driving sounds and other user-designed sounds, e. g. warning signals. The software was enhanced by a customized vehicle control module for highly automated driving and for cooperative decision making based on the four decision making automation designs.

The visual part of the CMDI was displayed on the middle screen as a head-up display (see Figure 5.6) and consisted of the following components:

- The available maneuvers, i. e. directions, at the next intersection, indicated by icons displaying respective arrows. The icons' background colors indicated the maneuvers' current status: gray indicated the non-availability of maneuver options, light blue indicated their availability; orange signaled the automation's choice of maneuver (and history), dark blue signaled the maneuver choice (history) of the driver; and green informed about an agreement on the corresponding maneuver.
- A countdown of 3 s, motivated by previous experiences [RWIH20] and displayed by means of a series of yellow triangles with respective numbers, before a cooperative decision making was enabled.

- A bar graph with a red background and a black rectangle, the size of which corresponded to the remaining time until the predefined deadline in a period of cooperative decision making had been reached.

Additionally and only for experimental design reasons, the objective measure of cooperative performance, associated with the travel time and explained further in the following, was displayed outside of the CMDI on the middle screen's top left corner, as well as a map of the overall Manhattan grid in the top right corner (see Figure 5.5 and 5.6). The display of the objective measure allowed participants to instantly assess the cooperative performance. The map showed the current position of the vehicle and the destination but no other traffic.

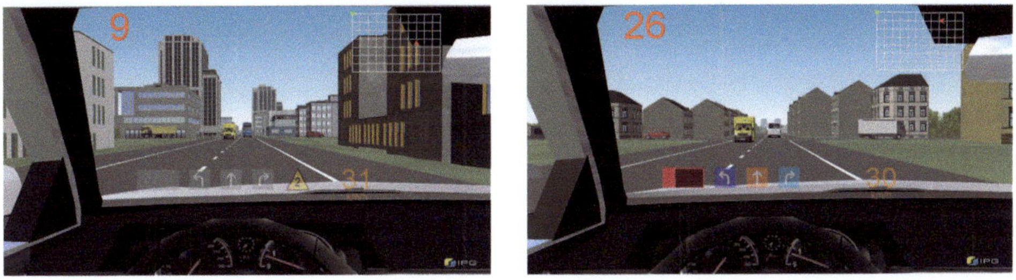

(a) Countdown phase prior to a cooperative decision making phase with disabled bar graph and decision options in gray color.

(b) Situation in a cooperative decision making phase with one maneuver choice of a participant (straight ahead, orange color) and the automation (left, dark blue color), the not chosen but available maneuver (right, light blue) and the bar graph (red & black).

Figure 5.6: Exemplary screenshots of the driving simulator's middle screen including the head-up display containing the display of a cooperative performance measure (top left), the available decision options i. e. maneuvers (center), the countdown display (right of center), a bar graph indicating the remaining time until the deadline (left of center) and the current vehicle speed (far right of center). ©2022 IEEE

Decision Scenarios

In general, decision scenarios comprise a set of decision options that cooperation partners are able to evaluate individually. If cooperation partners have to cooperatively decide on one decision option the following types of decision scenarios are possible:

- Conflict: In this scenario type, both cooperation partners have strong opposing preferences on the choice of a specific decision option. Hence, cooperation partners are forced into a cooperative decision making process to mutually decide for one decision option.
- Persuasion: In these scenarios, one cooperation partner is almost indifferent towards the decision options while the other cooperation partner has a strong

preference and therefore is expected to try to persuade the other cooperation partner.

- Trivial: Both cooperation partners prefer the same decision option and no process to reach an agreement is required.

A potential local traffic delay based on different types of vehicles (i. e. car, van, bus, truck) causing different but known delays was associated with each maneuver option. These delays can be contrasted to the time that it took to travel from one intersection to the next one without any traffic delays which was 14.5 s: car +3.6 s (+25 %), van +7.3 s (+50 %), bus +14.5 s (+100 %), truck +29 s (+200 %). In the following, a delay step or travel time step is defined for reasons of simplicity and readability as 3.6 s, i. e. the delay of a car.

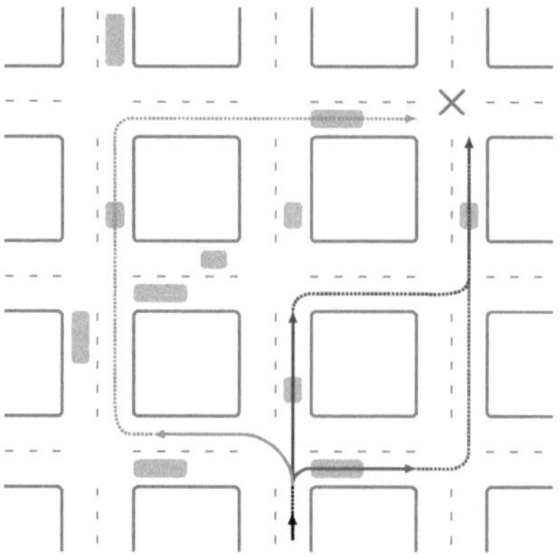

Figure 5.7: Exemplary segment of the Manhattan grid indicating traffic delays by gray rectangles (lengths represent the delay duration) and presenting the three decision options, i. e. maneuver options, for one decision scenario at the corresponding intersection by solid colored arrows. Respective optimal future paths to the destination (×) are depicted with dotted lines. ©2022 IEEE

While driving through the Manhattan grid, human participants and the automation were aware of their current position and destination by means of the displayed map (see Figure 5.6). Participants were also aware of the local traffic when approaching the intersection. Hence, they were able to assess the associated local delays. The automation had global information about the general traffic delays at all subsequent intersections (motivated by state-of-the-art real-time traffic information distribution and future car-to-x technology), yet it might have had false information about the local traffic delays at the next intersection (simulating the environment perception of the automation that requires some time for local information updating). The automation was therefore able to evaluate the globally required time to reach the destination for each decision option, yet potentially considering inaccurate local delays at the current intersection.

This setting emphasized the strength of both cooperating partners: The automated vehicle was well informed regarding the traffic along the upcoming route but could be tainted by potentially misinterpreted delays due to changing local traffic. The human was not able to anticipate future traffic but to perceive local traffic information correctly.

Consequently, local delays as well as misinformation were purposefully applied in the design of the Manhattan grid to create different maneuver preferences for human participants and the automation at each intersection, yielding the following instantiations of the different types of decision scenarios:

- Conflict: There were at least three delay steps between first and second maneuver preference for each cooperation partner and cooperation partners disagreed on first and second preferences.
- Persuasion: For one cooperation partner there was only one delay step in between the first and second maneuver preference, while the other cooperation partner had a strong preference, i. e. at least 3 delay steps in between.
- Trivial: Both cooperation partners evaluated the same maneuver option as the best. This type of scenario was applied to show the human the potential immediate agreement and that there was not always a conflict or persuasion situation.

The overall size of the Manhattan grid was 12×8 intersections consisting of 29 conflict scenarios, 33 persuasion scenarios and 30 trivial scenarios, disregarding the grid's corners. The detailed distribution of the scenario types in the Manhattan grid can be found in Table 5.2. Furthermore, the Manhattan grid is schematically depicted in Figure 5.8. The start position of the automated vehicle and the destination were placed on opposite corners of the grid. The globally optimal path to reach the destination without misinformation consisted of 6 conflict scenarios, 8 persuasion scenarios and 3 trivial scenarios. On this optimal path, traffic delays accumulated to 29 steps which was used as a baseline to compare the performance of the four different automation design to.

Each decision scenario started with a displayed countdown of 3 s. Within this time period the human cooperation partner was able to perceive the local traffic information regarding the upcoming intersection and the vehicle's position on the map. After the countdown, the actual phase of cooperative decision making started with the human cooperation partner being asked to communicate her or his most preferred option first. Afterwards, the automation would instantly present its most preferred option. After this, both cooperation partners were able to freely propose, i. e. select, other maneuver options without any regard of sequence nor fixed timing. The design of the beginning of the cooperative decision making process encouraged human attendance right from the start of the process. Hence, situations in which humans only react shortly before the deadline and do not take part in the decision

Table 5.2: Differentiation and distribution of scenario types in the Manhattan grid: total count within the Manhattan grid and count on the globally optimal path to reach the destination without misinformation.

Scenario Type	Scenario Name	Misinformation Present	Miscellaneous	Total Count	Count on Opt. Route
Conflict	S1	✘		15	2
	S2	✔		14	4
Persuasion	S3	✘	$\mathcal{A}$ persuades $\mathcal{H}$	16	4
	S4	✔	$\mathcal{A}$ persuades $\mathcal{H}$	7	2
	S5	✘	$\mathcal{H}$ persuades $\mathcal{A}$	10	2
Trivial	S6	✘		16	2
	S7	✘	on grid borders	14	1

making process were avoided, see insights of the suitability study reported on in Section 4.1. Therefore, this design was primarily a means for this experiment evaluating the cooperative decision making process. In other applications, designs in which the automation proposes first may be preferable.

Depending on how strong or weak the individual preferences (depending on the individual information on the difference of delay steps between different maneuver options) were, the automation designs based on the cooperative decision making models and/or the participants were expected to concede after some time (and potentially some decision option offering iterations): they were assumed to select additional maneuver options and hence agree with the cooperation partner on a maneuver choice. In case of the LA automation design or stubborn human behavior no agreement might have been reached. Then the ultimate decision was set according to the current automation design, i. e. automation choice in case of GT & LA and human choice in case of NT & LH. This reflected how the newly proposed automation designs try to close the gap between the two extremes in terms of authority assignment (LH & LA), as explained in Section 3.1.5. Hence, the phase of cooperative decision making ended either by an agreement on one maneuver option or by reaching the predefined deadline, i. e. the vehicle entering the intersection, after 9 s. This time was motivated by the assumption of at most three choices with 3 s each, as already applied in the models' suitability study, see Sections 4.1 and 4.2.1. The remaining time until reaching the deadline and entering the intersection was displayed by means of the bar graph for more clarity. After the deadline was reached, the resulting maneuver option as well as the current, updated measure of cooperative performance and its potential increase were displayed. The increase described the potentially added travel time steps of the resulting option with respect to the optimal path from the current intersection to the destination. Furthermore, the participant actually experienced the potential local traffic delay because the automated vehicle was slowed

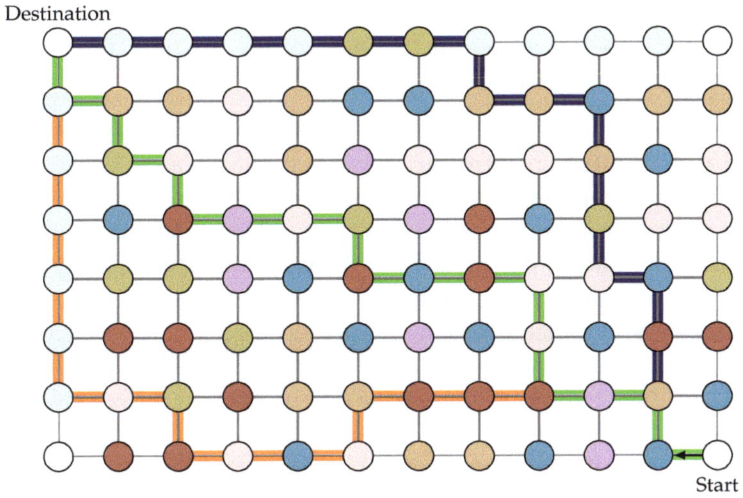

Figure 5.8: Schematic of the Manhattan grid: each circle and connecting line denote an intersection and a connecting street, respectively. The nodes' colors indicate the type of scenario when approaching this intersection while traveling towards the destination: S1, S2, S3, S4, S5, S6, S7, see Table 5.2. Three different important paths are the globally optimal path to the destination without misinformation (▬), the path considering only local information (▬), and the path considering global (mis-)information (▬).

down depending on the traffic associated with the conducted maneuver. This traffic disappeared before the next decision scenario started.

Automation Design

As already mentioned, four automation designs were evaluated in the course of the experiment: LH, LA, NT, and GT. All of these automation designs made their decisions based on the global and potentially on inaccurate local traffic delay information for each available direction of a given decision scenario introduced above.

In case of the automation design putting the human in the lead, i. e. LH, the automation might have proposed an own decision option but would ultimately accept the human decision without any resistance. In case the automation was in the lead, i. e. LA, the human might have proposed other decision options but the automation would ultimately follow through with its decision. By means of these behaviors, these automation designs followed the two potential manifestations of the leader-follower paradigm. Note that the application of decision support systems and dynamic role assignment approaches (see Section 2.3.2) was unrewarding in the considered decision scenarios: the scenarios were not as unclear such that a decision support would have been effective nor was a human intention identification for dynamically adapting the automation's authority rewarding due to the rather short decision making processes and potentially inaccurate information for the automation. Hence, LH and LA represent the state of the art with respect to cooperative decision making in these decision scenarios.

The automation designs NT and GT were based the adaptive negotiation model and the n-stage war of attrition introduced in Sections 3.2 and 3.3, respectively. They were designed and implemented in accordance with the guidelines of model-based automation design proposed in Section 4.2. As a result, the automation designs were capable of actually taking part in the cooperative decision making process with the human, i. e. the automation did not only display suggestions but also exhibited concession behavior in conflict situations. Furthermore, this concession behavior was human-like and its extent differed with respect to the model the automation designs were based on: the negotiation-theory-based automation design (NT) would give in as a last resort whereas the game-theory-based automation design (GT) ultimately insisted and realized its decision in case no agreement had been reached. The basis of the concession behavior of both NT and GT was the utility of the available decision options. These utilities were derived from the local and global delay information of maneuver options. To account for differences regarding the maximum and minimum delays of available maneuver options at different intersections, i. e. decision scenarios, data of each decision scenario were normalized. Refer to Appendix D.2 for more details on the model-based automation designs and parameterization.

Procedure

The overall practical accomplishment of the experiment took between 45 and 60 min and followed the procedure listed below.

1) **Introduction and Preparations**
Participants first read the guidelines on how to conduct the experiment. They were informed about the setup of the decision scenarios, i. e. explaining the Manhattan grid with intersections consisting of (usually) three decision options, the delays caused by the different types of vehicles at the intersections and the time to deadline. In addition, they were informed that the automation selects maneuver options based on information about additional delays at subsequent intersections and potentially false information about local delays. The objective for the participants was to reach a marked destination in the shortest possible time by iteratively and cooperatively deciding on a travel route. In each of the following experimental runs, they were unaware of the type of automation design, i. e. the exact maneuver-choosing behavior of the automation. Finally, the participants were asked to fill out the part of the custom-designed questionnaire (see Appendix D.3) regarding their general information and the familiarization procedure started.

2) **Familiarization Procedure**
   To introduce the general procedure of different decision scenarios and the handling of the decision interface, the participants were facing a shortened Manhattan grid (6×8) which consisted of random combinations of decision scenarios and automation designs. The results of this part were not included in the evaluation.

3) **First to Fourth Experimental Run**
   For each of the four automation designs the participants were passing one experimental run. The order of experimental runs were counterbalanced over participants to equate potential learning effects. Each experimental run was evaluated by the participants via a specific section of the custom-designed questionnaire. This scheme was applied to strengthen their sensitization and contemplation regarding the different automation designs.

4) **Postprocessing**
   After completing the fourth experimental run, the participants were asked to fill out the last part of the given questionnaire which allowed for an evaluation of the four experimental runs in relation to each other.

Participants

33 participants (27 male and 6 female) took part in the experiment. The average age was 29 years with an age range of 22 to 57 years. All participants possessed a valid driving license and 30.3 % did have some general experience regarding driving simulators.

Measures

The relevant measures for this experiment were an objective cooperative performance measure and subjective assessment by means of a questionnaire to evaluate the four experimental runs: Generally, the two automation designs based on the cooperative decision making models were compared with the two automation designs following the leader-follower approach. Furthermore, the relation of all four automation designs to each other was analyzed.

The objective cooperative performance regarding the human-machine cooperative decision making was measured by the additional travel time steps when comparing the required travel time at the end of each experimental run to the optimal route's travel time. Hence, the smaller the additional travel time steps, the higher was the performance of the human-machine cooperation.

To assess the participants' subjective perception of the human-machine cooperation, a questionnaire with a five-point *Likert Scale* [Lik32] with the following relevant items was used.

Q1: *How do you assess the overall cooperation between you and your partner?*
Answer range from *not satisfying* (1) to *satisfying* (5).

Q2: *How do you assess your partner's cooperation behavior?*
Answer range from *not reliable* (1) to *reliable* (5).

Q3: *How do you assess the interaction between you and your partner?*
Answer range from *not intuitive* (1) to *intuitive* (5).

Q4: *Was the behavior of your partner in cooperative decision making transparent?*
Answer range from *not transparent* (1) to *transparent* (5).

The entire questionnaire is provided in the Appendix D.3.

Due to the comparison of up to four sample sets and the lack of information regarding their distributions, the statistical analysis was conducted by means of the non-parametric *Kruskal-Wallis* test [KW52]. The test's null hypothesis (all sample sets origin from the same original distribution) was accepted if $\mathcal{H} \leq \chi_{\mathrm{c}}^2$ holds. In case of the pooled comparison of the two automation designs based on the state-of-the-art *leader-follower* models (LH & LA) with the two newly-introduced automation designs based on the cooperative decision making models (GT & NT) $\chi_{\mathrm{c}}^2 = \chi_{df=1,\alpha=0.05}^2 = 3.842$ follows. When comparing the individual results of the four automation designs, there were three degrees of freedom ($df = 3$). Hence, with a significance level of $\alpha = 0.05$, $\chi_{\mathrm{c}}^2 = \chi_{df=3,\alpha=0.05}^2 = 7.815$ follows.

Based on these measures, the following section provides the results of the conducted experiment.

5.3.2 Results

First, objective performance results are provided to investigate Hypothesis 5.2. Figure 5.9 shows the objective cooperative performance by means of compact boxplots (see explanation in Appendix D.1) based on the additional travel time steps for each automation design. It reveals that experimental runs with the automation designs based on cooperative decision making models yielded less additional time steps than the leader-follower-based automation designs. Furthermore, comparing the pooled automation designs LA & LH with the pooled automation designs GT & NT, the null hypothesis of the Kruskal-Wallis test was rejected with $\mathcal{H} = 72.123$. Considering the sample set for the four automation designs individually, the null hypothesis was rejected with $\mathcal{H} = 64.823$. Hence, the objective cooperative performance measure was significantly better for the automation designs based on cooperative decision making

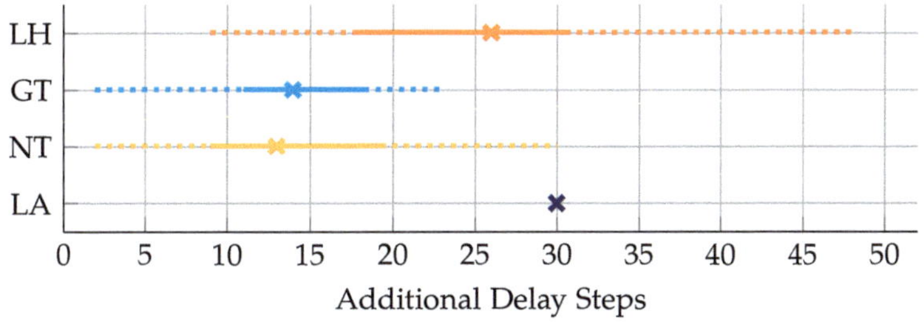

Figure 5.9: Compact boxplots (see explanation in Appendix D.1) of additional travel time steps for each automation design. Median ×, lower/upper quartile —, lower/upper adjacent · · · .

models than for the leader-follower automation designs. Therefore, Hypothesis 5.2 was accepted.

Next, the participants' subjective assessment is provided to investigate Hypothesis 5.3. Figure 5.10 shows the participants' subjective perceptions based on the corresponding questions Q1-Q4 of the questionnaire. Comparing the pooled automation designs LA & LH with the pooled automation designs GT & NT, the null hypothesis of the Kruskal-Wallis test was rejected regarding the satisfaction with the human-machine cooperation ($\mathcal{H} = 83.776$), the trust in automation's decision making behavior ($\mathcal{H} = 52.51$), the intuition of the interaction ($\mathcal{H} = 24.192$) and the transparency of the interaction ($\mathcal{H} = 7.563$). In view of the individual sample sets of the four automation designs, the null hypothesis of the Kruskal-Wallis test was also rejected regarding the satisfaction with the human-machine cooperation ($\mathcal{H} = 84.845$), the trust in automation's decision making behavior ($\mathcal{H} = 52.682$), the intuition of the interaction ($\mathcal{H} = 24.85$) and the transparency of the interaction ($\mathcal{H} = 11.406$). To sum up, the evaluation of subjective perception regarding the different automation designs revealed that the automation designs based on cooperative decision models led to a significantly more satisfying, trustworthy and intuitive interaction in comparison to the state-of-the-art leader-follower approaches. However, the opposite held for the transparency of the interaction. Therefore, Hypothesis 5.3 was accepted.

In summary, both hypotheses stated at the beginning of Section 5.3 were accepted.

For a deeper understanding, some post-test results for each measure comparing the sample sets of each automation design individually by means of a t-test are provided. All resulting p-values are given in Table 5.3. Considering the objective cooperation performance measure, all sample sets differed significantly except for the comparison of NT & GT. Regarding the participants' satisfaction with the human-machine cooperation, the trust in the automation's decision making behavior and the intuition of the interaction between human and automation, the sample sets of both NT and GT were significantly different compared to LH and LA. Considering the transparency of the interaction between human and automation, there were significant

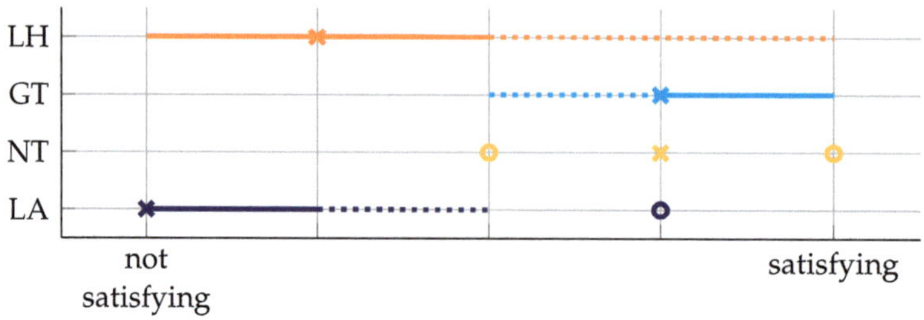

(a) Q1: satisfaction regarding the human-machine cooperation

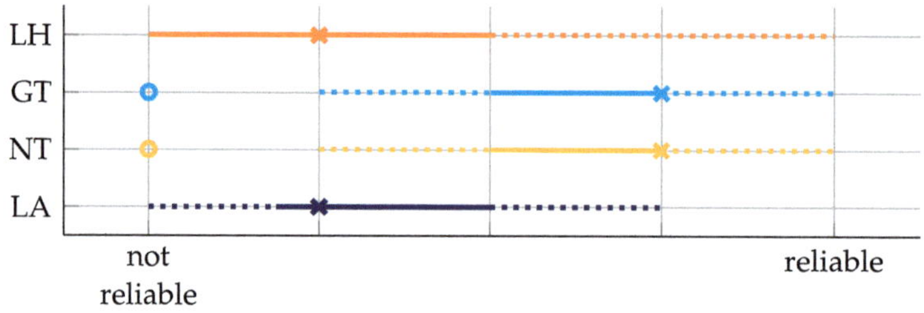

(b) Q2: trust in automation's decision making behavior

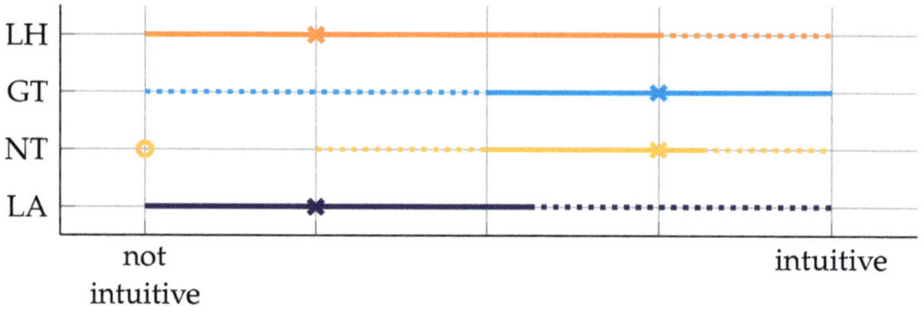

(c) Q3: intuition of the interaction

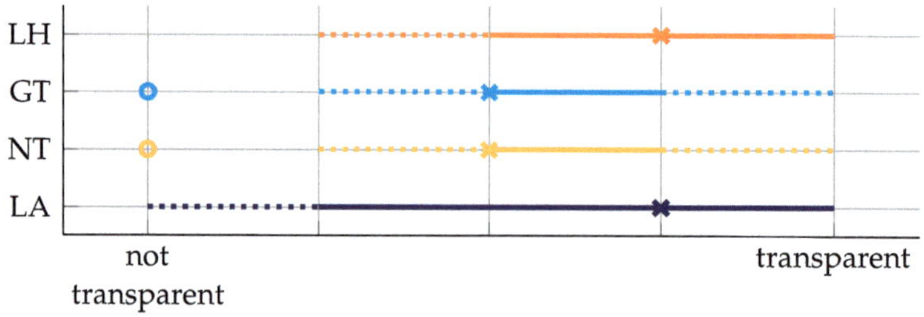

(d) Q4: transparency of the interaction

Figure 5.10: Compact boxplots (see explanation in Appendix D.1) regarding the subjective perceptions to Q1-4. Median ×, lower/upper quartile —, lower/upper adjacent · · ·, and outlier o.

Table 5.3: Results of the t-test evaluating objective performance measure and answers to Q1-Q4: p-values of pair-wise comparison.

Measures \ Pairs	LA NT	LA GT	LA LH	NT GT	NT LH	GT LH
Obj. Coop. Performance	0	0	0.042	1	0	0
Q1: Satisfaction	0	0	1	1	0	0
Q2: Trust	0	0	1	1	0	0
Q3: Intuition	0.004	0	1	1	0.022	0.003
Q4: Transparency	1	1	0.304	1	0.016	0.026

differences comparing the sample sets of GT and NT with LH and no significant difference in comparison with LA.

Furthermore, the objective cooperation performance measure strongly correlated to participants' subjective assessment of the satisfaction with the human-machine cooperation ($M = -0.8113$, $SD = 0.2195$). In other words, participants were more satisfied with the human-machine cooperation if the cooperation led to smaller travel times (a better performance), and vice-versa.

The above gained insights were also supported by collected statements of participants noticing a "will to compromise" and "good proposals" of the automation designs based in the cooperative decision making models. The interaction with them was perceived as "pleasant" and "trustworthy". The interaction with leader-follower approaches was criticized as "frustrating" and "strenuous". Participants perceived the automation design with the automation in lead as "too dominant" and "unresponsive to suggestions". When participants were in lead the automation was criticized for "taking no responsibility".

5.3.3 Discussion

The significantly improved objective cooperative performance for the automation designs based on cooperative decision making models compared to the leader-follower automation designs demonstrates that

- an emancipated design of the human-machine cooperation on decision level is beneficial for the overall cooperative system's performance and that
- a model-based approach is suitable to design the corresponding automation.

Furthermore, note that the objective performances of LA and LH did not differ significantly, i. e. the performance of LA was reasonably designed and did not systematically void the results.

The observed significantly more satisfying and intuitive interaction with the automation designs based on cooperative decision models may have been a result of the significantly increased trust regarding the automation's decision making behavior. In other words, participants recognized the increased cooperative, i. e. concessive, behavior of the introduced cooperative decision model automation designs as more trustworthy and intuitive which also increases participants' acceptance of the automation.

A closer look at the reduced transparency of interaction for the automation designs based on cooperative decision making models reveals two insights:

1. Designing fully automated systems is not necessarily the solution in terms of transparency for humans as interaction with LA was not assessed significantly more transparent than GT & NT.
2. Humans prefer complete transparency about the final decision as interaction with LH is assessed significantly more transparent compared to all other automation designs which was expected as the human has exclusive control authority. Hence, in terms of transparency, assistive decision support systems have an advantage compared to emancipated decision making system.

Putting together all these insights, the trade-off in designing cooperative systems becomes apparent, i. e. balancing the aspects of cooperative performance, human acceptance, trust in the automation, intuition and transparency of interaction. According to the experiment's results and depending on the application context, approaches with focus on cooperative decision making or humans in lead are preferable in contrast to approaches with the automation in lead.

5.3.4 Conclusion

This experiment yielded results which demonstrate that the proposed automation designs for cooperative decision making based on negotiation theory and game theory add value for human-machine cooperation on decision level in the examined scope of highly automated vehicles: the objective cooperative performance was significantly increased compared to automation designs based on conventional leader-follower approaches. While the transparency of interaction slightly decreased as expected, the remaining aspects of the subjective assessment of the participants in terms of satisfaction and trust in the cooperation as well as intuition of interaction revealed a preference for cooperative decisions models. This reveals the known trade-off in cooperative system design to accommodate the increased cooperative performance, human acceptance of and trust in the automation, and the transparency of interaction.

To summarize, the experiment evidently reveals humans' preference for an emancipated interaction on decision level.

5.4 Conclusion of the Experimental Evaluation

Both reported experiments pursued the general evaluation approach introduced in Section 5.1 and therefore provided first empirical evidence that cooperative performance is significantly increased by allowing for emancipated human-machine cooperative decision making. Furthermore, the subjective evaluation reveals that humans prefer this truly cooperative interaction over state-of-the-art leader-follower approaches in terms of user acceptance of and trust in the automation. The mixed subjective assessments with respect to intuition and transparency of the interaction demonstrate the relevance of finding a trade-off in the design of cooperative systems, i. e. finding the balance between increased cooperative performance and subjective human assessment of not being in full control.

Consequently, the two experimental evaluations demonstrate in realistic simulations the ability of enabled automation designs to cooperatively and effectively make decisions with humans. Furthermore, the proposed mathematical behavior models of human-machine cooperative decision making and corresponding automation designs successfully close the gap between fully automated and human-centered decision making from a practical point of view (see Section 3.1.5) and answer the third research question of this thesis, see Section 2.4.

Additionally, the newly gained insights add major value for the design of future cooperative systems by expanding their widespread practical limitation to the action level of human-machine cooperation towards explicitly including the decision level. Hence, the experimental results revealing the benefits of emancipated human-machine cooperation on decision level encourage further research and practical applications.

6 Conclusion

This thesis focuses on the decision making aspect of human-machine cooperation: It provides evidence that *emancipated human-machine cooperative decision making outperforms human individualism and technical autonomy* in terms of objective performance, user satisfaction, and human trust in the interaction.

Along the way to this novel insight into cooperative human-machine systems' design, this thesis initially analyzes the current state of research on human-machine cooperation and proposes the *butterfly model* as a comprehensive classification of human-machine cooperation. On this basis, the research gap on the decision level of human-machine cooperation is revealed: there is no approach reported in literature that enables the machine to take part in an *emancipated* human-machine cooperative decision making process, i. e. human and machine participate in a process of cooperative decision making with equal authority.

To close this gap, this thesis subsequently proposes a first *meta-model of emancipated human-machine cooperative decision making*. This meta-model takes into account the human limitations and characteristics in a cooperative decision making scenario. Applying this meta-model as a design template, this thesis introduces two mathematical behavior models for emancipated human-machine cooperative decision making processes: the *adaptive negotiation model* and the *n-stage war of attrition* which originate from negotiation theory and game theory, respectively. In case of the adaptive negotiation model, the cooperative decision making process modeling is inspired by negotiating automated, i. e. programmable, agents whereas in case of the n-stage war of attrition the focus is on selfish rational entities, e. g. humans. In both cases, a concessive process of exchanging decision option offers is established to the end of reaching a mutual agreement. Furthermore, both models account for the uncertainty in cooperative decision making with human participation: The adaptive negotiation model provides the ability to identify the negotiation behavior of the cooperation partner and adapt the own behavior accordingly. The n-stage war of attrition inherently considers uncertainty and allows for an adaptation of the interaction strategy based on observed actions of the cooperation partner. In decision making scenarios with a given deadline, the adaptive negotiation model furthermore provides a theoretical guarantee for reaching a mutual agreement. In contrast to this, the n-stage war of attrition only considers soft deadlines which in turn allows for emulating unyielding behavior. As a result, the two mathematical behavior models successfully close the gap between the two extremes of the state-of-the-art leader-follower approach, i. e. the human or (more rarely) the automation being in the lead, towards an emancipated human-machine cooperation.

For the purpose of experimentally investigating both models of human-machine cooperative decision making, this thesis reports on a study and corresponding results *prove the suitability* of the basic negotiation model and the n-stage war of attrition to *describe human concession behavior*. Furthermore, the study's results highlight the necessity of an adequate interface design for cooperative decision making. Encouraged by the study's results, *two automation designs based on the two proposed mathematical behavior models of cooperative decision making* are introduced along with guidelines for their practical implementation. By means of the mathematical behavior models' ability to represent human concession behavior, the automation designs additionally aim for an intuitive human-machine cooperation and high user acceptance. A potential preference for the application of one of the proposed automation designs depends on the application scenario and the features of the respective mathematical model of human-machine cooperative decision making.

Pursuing the empirical evidence for the benefits of emancipated human-machine cooperative decision making, this thesis proposes a novel experimental design by introducing specific requirements and measures for subjective and objective cooperative performance evaluation focusing on the decision making aspect of human-machine cooperation. Following these guidelines, this thesis reports on *two experimental evaluations* of the newly proposed automation designs. The first experiment's scope is the cooperative determination of the appropriate LOA in *teleoperating a mobile robot* in a search-and-rescue scenario. The other experiment is set in the scenario of *highly automated driving* in which the driver and the vehicle's automation have to cooperatively decide on the selection of driving maneuvers. In both experiments, the proposed automation designs were compared to state-of-the-art approaches. The results demonstrate the benefits of the novel automation designs capable of emancipated human-machine cooperative decision making in terms of objective cooperative performance and subjective user satisfaction and trust in the cooperative systems. Hence, both experiments provide first evidence that humans prefer an emancipated cooperation on decision level. Furthermore, performance benefits can be created or increased by considering this form of cooperation. Therefore, it can be concluded that emancipated human-machine cooperation on decision level has the ability to outperform the individual decision making of either human or automated system and raises synergies from both perspectives of objective system design and subjective user perception.

These novel positive insights into the research on human-machine cooperation may encourage further research on emancipated human-machine cooperative decision making. The experimental results highlight the necessity to further elaborate the interface design for cooperative decision making. Additionally, the application of the automation designs to other fields of human-machine cooperative decision making has to be investigated in order to explore novel scopes and also potential practical limitations.

Another major challenge remaining with respect to the cooperative human-machine system design is the seamless shift of human-machine cooperation across all levels of task abstraction. Hence, extensive research is required which enhances existing approaches on the action level of human-machine cooperation by means of the proposed approaches on decision level.

Therefore, this thesis advances research towards the ultimate goal in cooperative systems' design which is a holistic consideration and realization of human-machine cooperation on all levels of task abstraction and with a large area of application. Regarding the disadvantages of fully automated systems in terms of high development costs and out-of-the-loop problems for human supervisors, this research therefore strengthens the superior alternative, i. e. the application of cooperative human-machine systems.

A Mathematical Fundamentals

This appendix provides relevant mathematical fundamentals for more complex integration and differentiation as well as for the transformation of density functions.

A.1 Definition of Integrals with Infinite Integration Limits

Integrals with infinite integration limits are defined as follows.

Definition A.1 (Definition of Integrals with Infinite Upper Integration Limits)
Integrals with an infinite upper integration limit are defined as follows [BSMM15, p. 507]:

$$\int_a^\infty f(x)\,\mathrm{d}x = \lim_{b \to \infty} \int_a^b f(x)\,\mathrm{d}x \qquad a, b \in \mathbb{R}, a < b. \tag{A.1}$$

A.2 Differentiation for Limits of and Under the Symbol of Integrals

In order to differentiate limits or the integrand of an integral, the following differentiation rule applies.

Lemma A.1 (Differentiation for Limits of and Under the Symbol of Integrals)
Consider continuous, differentiable and bounded limit functions $\alpha(y)$ and $\beta(y)$ defined on a finite interval of y and a continuous integrand $f(x,y)$ with a continuous partial derivative with respect to y, then the following differentiation rule holds [BSMM15, p. 512]:

$$\frac{\mathrm{d}}{\mathrm{d}y}\int_{\alpha(y)}^{\beta(y)} f(x,y)\,\mathrm{d}x = \int_{\alpha(y)}^{\beta(y)} \frac{\partial f(x,y)}{\partial y}\,\mathrm{d}x + \frac{\mathrm{d}\beta(y)}{\mathrm{d}y}\cdot f(\beta(y)) - \frac{\mathrm{d}\alpha(y)}{\mathrm{d}y}\cdot f(\alpha(y)) \tag{A.2}$$

Proof:
Be referred to [BSMM15, p. 512]. □

A.3 Density Function Transformation

The following lemma provides the mathematical relation between a transformed density function and its original.

Lemma A.2 (Density Function Transformation)
Consider a one-dimensional density function $f_x(x)$ (non-negative and Lebesgue-integrable) and a scalar, invertible transformation $y = \phi(x), \phi : \mathbb{R} \mapsto \mathbb{R}$. The inverse transformation is denoted by ϕ^{-1}.
The transformed density function $f_y(y) = f_y(\phi(x))$ is given by:

$$f_y(y) = f_x\left(\phi^{-1}(y)\right)\left|\frac{\mathrm{d}}{\mathrm{d}y}\left(\phi^{-1}(y)\right)\right|. \tag{A.3}$$

Note that the transformation of the corresponding cumulative distribution function $F_x(x) = \int_{-\infty}^{x} f_x(\tilde{x})\,\mathrm{d}\tilde{x}$ by means of ϕ results in

$$F_y(y) = F_x\left(\phi^{-1}(y)\right). \tag{A.4}$$

Proof:
This transformation results from the *substitution method* [BSMM15, p. 484]. □

B Application Example of the Adaptive Negotiation Model

The following application example explores by simulating a human-machine negotiation the potential of the adaptive negotiation model in terms of negotiation behavior identification, adaptation towards the identified behavior. Furthermore, it demonstrates how offers can convey additional information for the cooperative decision making process besides the information about the associated decision options.

The exemplary application of the adaptive negotiation model is the negotiation of directions at an interaction between a highly automated vehicle and human driver. For the simulation of this scenario, both agents are modeled by means of the introduced adaptive negotiation model, see Section 3.2. Both agents are able to exchange offers which represent a proposed decision option and the (potentially time-variant) importance of that choice. In the following, the scenario and the agents' setup are presented in more detail before the simulation results are shown.

B.1 Scenario

The exemplary road scenario is a *Manhattan grid* navigation setting depicted in Figure B.1. The aim of both agents is to reach the intersection marked with a green dot. At the time of the negotiation the vehicle is traveling along the black solid arrow. At the intersection three decision options d are available for both agents: turn left (d^1), drive straight ahead (d^2) and turn right (d^3). Each decision option can be offered

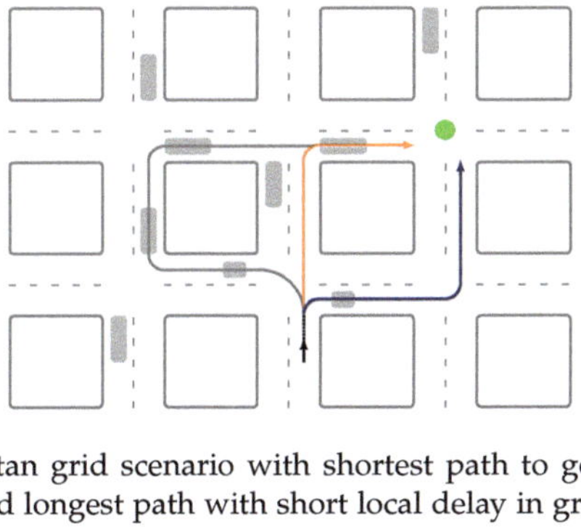

Figure B.1: Exemplary Manhattan grid scenario with shortest path to goal in blue, path avoiding local delays in orange and longest path with short local delay in gray.

with one of three importance levels $\zeta_i \in Z$, $|Z| = 3$. Consequently, an offer is described by the tuple $o := (d, \zeta)$ and the set of offers has a magnitude of $\|O\| = 9$. The importance levels represent an additional communication parameter that indicates how much an agent clings to the chosen direction with respect to the agent's concession strategy and the directions' utility differences. As the choice of importance level is influenced by the agent's time-based concession behavior, the other agent's identification of the agent's negotiation behavior is able to take into account this additional information and is hence facilitated and quicker.

In Figure B.1, the gray boxes indicate traffic delays. The options d can be rated with respect to to the time loss due to a local traffic delay t_l at the current intersection and to the estimated time to reach the target intersection t_g taking into account all relevant traffic delays on the remaining way. The simulation results for the proposed model are based on the times in Table B.1.

Table B.1: Times for local traffic delay and time to goal intersection.

D	t_g	t_l
d^1 (left)	390	10
d^2 (straight)	140	0
d^3 (right)	80	40

The negotiation is set to start at time $t = 0$ and agents face a deadline $t = \mathcal{T}$ at which the vehicle has to start one of the potential maneuvers. The time during the negotiation is normalized, i. e. $\bar{t} := t/\mathcal{T}$, $\bar{t} \in [0,1] \subset \mathbb{R}$.

B.2 Agents' Setup

Due to the introduction of additional communications symbols in form of importance levels, agents need to determine the importance level along with the direction to provide offers $o = (d, \zeta)$. Hence, the utility functions for both agents are set as a linear combination of evaluation functions for evaluating the decision option d and the importance level ζ of offer o:

$$u_i(o) = u_i(d,\zeta) := \underbrace{w_{g,i} \cdot \bar{b}_g(d) + w_{l,i} \cdot \bar{b}_l(d)}_{\bar{u}_i(d)} + b_\zeta(\zeta) \tag{B.1a}$$

with

$$\bar{b}_g(d) := \frac{\min_{\forall d^\mu \in D} t_g(d^\mu)}{t_g(d)}, \tag{B.1b}$$

$$\bar{b}_l(d) := 1 - \frac{t_l(d)}{\sum_{\forall d^\mu \in D} t_l(d^\mu)}, \tag{B.1c}$$

$$b_\zeta(\zeta) := \begin{cases} \underbrace{-0.5 \cdot \dfrac{\zeta - \min(Z)}{\max(Z) - \min(Z)}}_{\bar{b}_\zeta(\zeta)} & \text{if (B.1f) holds} \\ \infty & \text{else} \end{cases} \tag{B.1d}$$

s. t.

$$w_{g,i} + w_{l,i} = 1, \tag{B.1e}$$

$$\bar{b}_\zeta(\zeta) < \bar{u}_i(d) - \max_{d^\mu \in \tilde{D}} \bar{u}_i(d^\mu) \tag{B.1f}$$

$$\tilde{D} := \{d^\mu \in D | \bar{u}_i(d) > \bar{u}_i(d^\mu)\}. \tag{B.1g}$$

$\bar{b}_g(d)$ penalizes the time for reaching the target intersection, referred to as the time-to-goal t_g, of a decision option d with respect to the fastest alternative. $\bar{b}_l(d)$ penalizes the local traffic delay t_l of decision option d by comparing it to the sum of all local traffic delays. $\bar{b}_\zeta(\zeta)$ penalizes the usage of importance levels for communication. This models the importance level as a measure for the deviation of the utility of the chosen direction $\bar{u}_i(d)$ from the target utility $\bar{u}_{t,i}$. The agents will start with minimum importance level, increase it when approaching the next closest utility of another direction and restarting with minimum level of importance whenever offering a new decision option. The cases in (B.1d) with condition (B.1f) ensure that higher importance levels are only communicated in case their associated decision option is still valid, i. e. no other offer comprising another decision option has been proposed since this associated decision option has been offered. Therefore, note that in (B.1a) $u_i(o) \in [0,1] \cup \infty$. However, this does not negatively influence the concession strategy: the optimal offer $o_i^t = (d^*, \zeta^*)$ at time instance t is determined following the time-based concession strategy of Definition 3.8, i. e. solving the optimization problem (3.4) utilizing $u_i(o)$ defined in (B.1a).

For the simulation of the negotiation between agent $\mathcal{A}$, resembling the automation and focusing on the time to goal, and agent $\mathcal{H}$, the human, trying to avoid local traffic delays, the agents are parameterized as follows:

$$\epsilon_{\mathcal{A}/\mathcal{H}} = 1, \quad w_{g,\mathcal{A}} = 1, \quad w_{g,\mathcal{H}} = 0, \quad w_{l,\mathcal{A}} = 0 \quad w_{l,\mathcal{H}} = 1.$$

Both agents are able to identify the other agent's parameters $\boldsymbol{\theta}_j = [\epsilon_j, w_{g,j}]^\top$, $j \in \{\mathcal{A}, \mathcal{H}\}$, by means of the identification method presented in Section 3.2.4. In this setting, the following aspects of the identification method are adapted with respect to the introduced negotiation scenario: In order to calculate the Bayesian update, $p\left(o_j^t \mid \boldsymbol{h}_l\right)$ has to be determined. This likelihood depends on a-priori knowledge on the other agent's behavior and observed offers o_j^t and can be reformulated to:

$$\begin{aligned} p\left(o_j^t \mid \boldsymbol{h}_l\right) &= p\left(d_j^t, \zeta_j^t \mid \boldsymbol{h}_l\right) \\ &= \frac{p\left(d_j^t, \zeta_j^t, \boldsymbol{h}_l\right)}{p(\boldsymbol{h}_l)} \\ &= \frac{p\left(\zeta_j^t \mid d_j^t, \boldsymbol{h}_l\right) \cdot p\left(d_j^t \mid \boldsymbol{h}_l\right) \cdot p(\boldsymbol{h}_l)}{p(\boldsymbol{h}_l)} \\ &= p\left(\zeta_j^t \mid d_j^t, \boldsymbol{h}_l\right) \cdot p\left(d_j^t \mid \boldsymbol{h}_l\right). \end{aligned} \tag{B.2}$$

$p\left(d_j^t \mid \boldsymbol{h}_l\right)$ depends on the concession and acceptance strategy, i. e. (3.4) and (3.2), respectively. Therefore, the associated direction of offer o_j^t of the other agent has to fulfill the following condition:

$$\begin{aligned} \bar{u}_{\boldsymbol{h}_l}\left(d_j^t\right) &= \min_{d \in D} \bar{u}_{\boldsymbol{h}_l}(d) \\ \text{w. r. t.} \quad & \bar{u}_{\boldsymbol{h}_l}(d) \geq \bar{u}_{\mathrm{t},h}(t) \text{ and} \\ & \bar{u}_{\boldsymbol{h}_l}(d) > u_{\boldsymbol{h}_l}(d_i^t) \end{aligned} \tag{B.3}$$

The index $\square_{\boldsymbol{h}_l}$ indicates the parameterization of the corresponding function with the parameters of hypothesis $\boldsymbol{h}_l$. Besides ensuring that the other agent's utility of the chosen direction lies above target utility, condition (B.3) also takes into account that this utility must be higher than that of the last own offer with respect to the other agent's utility measure. Otherwise this offer would have been accepted by the other agent.

All hypotheses fulfilling this condition explain the current chosen direction of the other agent. Therefore a uniform distribution is assigned to these hypotheses:

$$p\left(d_j^t \mid \boldsymbol{h}_l\right) := \begin{cases} \frac{1}{|\tilde{D}|} & \text{if (B.3) holds} \\ 0 & \text{else} \end{cases} \tag{B.4}$$

$$\text{with} \qquad \tilde{D} := \{d \in D \mid \text{(B.3) holds}\}.$$

Note that in this exemplarily case $\tilde{D}$ is a singleton.

The probability $p\left(\zeta_j^t \mid d_j^t, \boldsymbol{h}_l\right)$ of an importance level ζ_j^t given a direction d_j^t and a parameterization $\boldsymbol{h}_l$ depends on the concession strategy (3.4) with respect to (B.1a). Therefore the following condition has to hold:

$$\zeta_j^t = \arg\min_{\zeta \in Z} \left\{ u_{\boldsymbol{h}_l}\left(d_j^t, \zeta\right) - \bar{u}_{\mathrm{t},\boldsymbol{h}_l}(t) \right\} \tag{B.5}$$

All hypotheses that fulfill this condition explain the current chosen importance level at the current direction. Due to the fact that only one importance level per direction is valid, the probability is set to

$$p\left(\zeta_j^t \mid d_j^t, \boldsymbol{h}_l\right) := \begin{cases} 1 & \text{if (B.5) holds} \\ 0 & \text{else} \end{cases} \tag{B.6}$$

Furthermore, the probability re-initialization offset is set to $q = 0.001$. Aside from that, agent $\mathcal{A}$ is able to adapt its negotiation behavior with $\beta = 0.8$ and $r_{\mathcal{A}} = 0.3$, see Section 3.2.5. Moreover, agent $\mathcal{A}$ is set to propose offers at a constant update rate whereas agent $\mathcal{H}$, representing the human, interacts at random times.

B.3 Simulated Negotiation Process

Figure B.2 shows a negotiation process without adaptation. The agreement on option d^2 is indicated by a green circle. The vertical bars represent different levels of importance. Note that due to the asynchronous protocol the agents are allowed to interact at random times. Therefore, agent $\mathcal{H}$ detects the agreement only at his next interaction time. The corresponding performance of the identification method of agent $\mathcal{A}$ is depicted in Figure B.3. The estimated values (dashed lines) converge from their starting values at $\bar{t} = 0$ towards the real values (solid line). Note that changes in direction offered or in importance levels contribute most to improvements regarding the parameter estimation, as they provide a high information content.

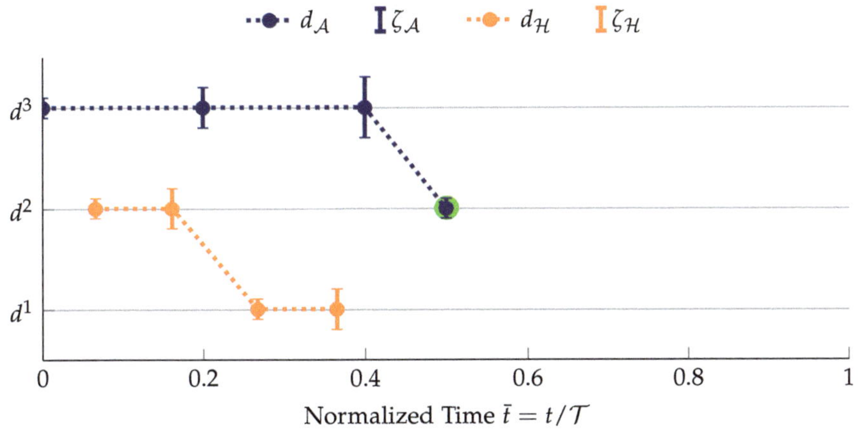

Figure B.2: Negotiation process without adaptation: green circle indicates agreement.

Figure B.4 shows a negotiation round in which agent $\mathcal{A}$ adapts its behavior after the identification process of the agent $\mathcal{H}$ model's parameters is about to converge.

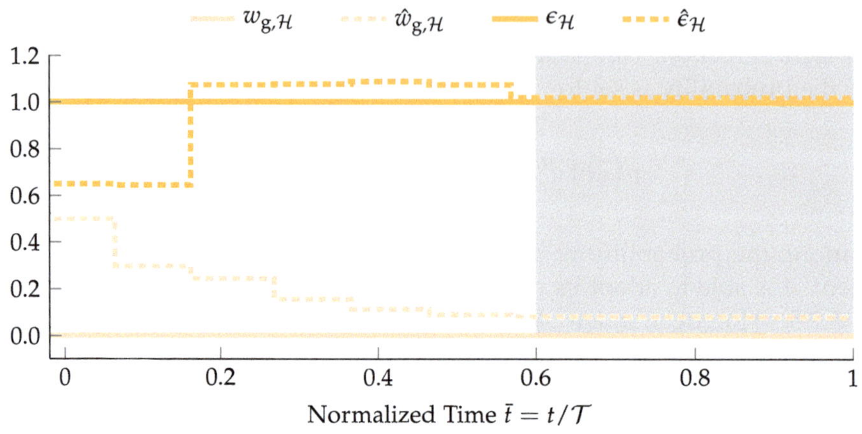

Figure B.3: Identification process of agent $\mathcal{A}$ without adaptation. Actual parameters are depicted with solid lines, dotted lines represent the estimates.

Agent $\mathcal{A}$ becomes more intransigent and therefore is able to convince agent $\mathcal{H}$ with his offer for option d^3. Figure B.5 presents the identification performance of agent $\mathcal{H}$ of the changing behavior of agent $\mathcal{A}$. The adaptation process is visible regarding the changing blue trajectories of the concession parameter $\epsilon_{\mathcal{A}}$ from high to low values, i. e. from concessive to intransigent behavior. Also the identification ability of changing negotiation behavior is visible as the estimates follow the actual values with a small delay.

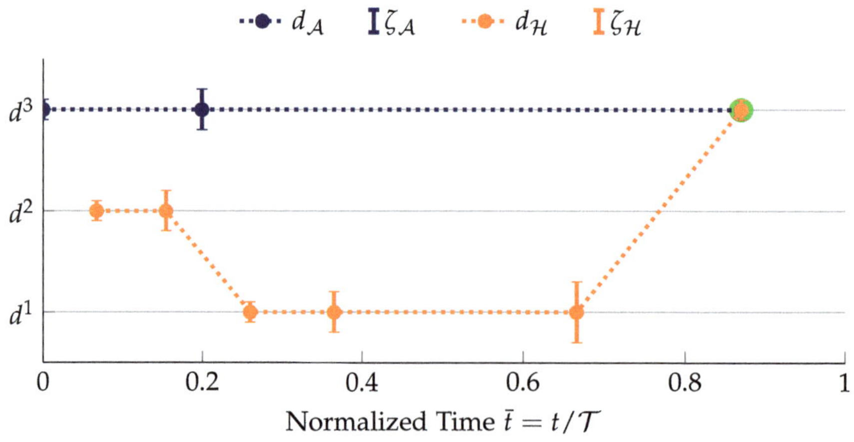

Figure B.4: Negotiation process with adaptation: green circle indicates agreement.

In conclusion, the simulated adaptive model is able to model negotiation scenarios that lead to an agreement between emancipated agents. The agents are allowed to

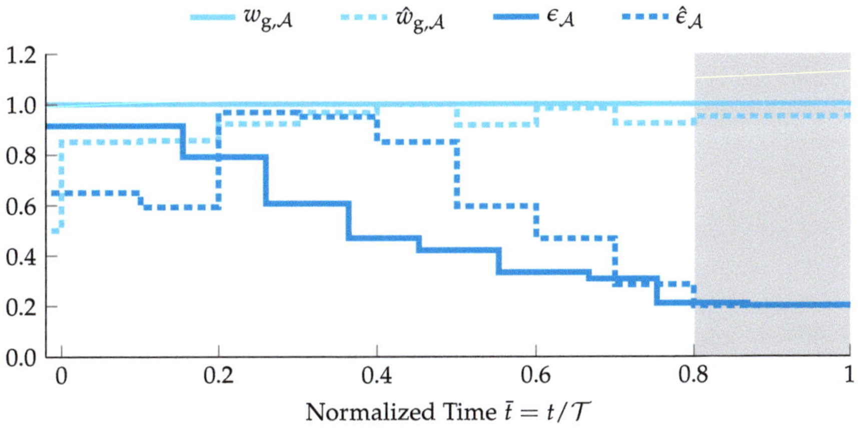

Figure B.5: Identification process of agent $\mathcal{H}$ showing adaptation of agent $\mathcal{A}$. Actual parameters are depicted with solid lines, dotted lines represent the estimates.

communicate at different rates and with importance levels as additional communication symbols. Furthermore, the proposed identification method is able to identify the behavior of the other agent (see Figure B.3), even if it is changing, see Figure B.5. The explicit adaptation strategy allows the agent to change his negotiation behavior based on the estimated effort and outcome of persuading the other agent, see Figure B.4. As a result the outcome of the negotiation may be different to the one without adaptation. The ability to adapt with respect to some objective function, in this case the trade-off between outcome utility and effort to achieve it, is a great advantage of the introduced model. In comparison to existing adaptation techniques, the introduced approach is more generalized and allows for more efficient negotiations.

C Supplementals on Game Theory

This appendix provides some important supplementals on game theory for this thesis. It states the definitions of important equilibria followed by an additional lemma on the sufficiency of a condition on the maximum payoff of the applied war of attrition game model.

C.1 Important Equilibria

Equilibria in games define the state of strategy profiles. In the following, equilibria definitions are provided for games with two players. The most famous equilibrium for complete information games is the *Nash equilibrium*.

> **Definition C.1 (Nash Equilibrium for Two Players)**
> *Consider a strategy profile* $\left(\psi_i^*, \psi_j^*\right)$, $i, j \in P, i \neq j$ *in a complete information game. The profile is in a Nash equilibrium if the following inequality condition for the payoff holds for all players:*
>
> $$\pi_i\left(\psi_i^*, \psi_j^*\right) \geq \pi_i\left(\psi_i, \psi_j^*\right), \ \forall \psi_i \in \Psi_i, \forall i \in P. \tag{C.1}$$
>
> *A strict Nash equilibrium is given if*
>
> $$\pi_i\left(\psi_i^*, \psi_j^*\right) > \pi_i\left(\psi_i, \psi_j^*\right), \ \forall \psi_i \in \Psi_i, \forall i \in P. \tag{C.2}$$
>
> *(see Definition 1.2 in [FT91, p. 11])*

In games with incomplete information, the analogue to the Nash equilibrium is the *Bayesian Nash equilibrium*. It incorporates the *type* of a player which resembles players' private information. This incomplete information about the other player usually considers the player's payoff which is why rational players choose strategies that maximize the *expected payoff* with respect to to a *belief* about the potential type of the other player. This belief depends on a common knowledge probability distribution of types and potentially also on the player's own type.

Definition C.2 (Bayesian Nash Equilibrium)
Suppose the strategies $\psi_i \in \Psi_i$ *of players* $i \in P$ *depends on their type* λ_i *which is private information. Furthermore, the type's probability density function* $f(\lambda_i, \lambda_j)$ *is given and common knowledge. The strategy profile* $\left(\psi_i^*(\lambda_i), \psi_j^*(\lambda_j)\right)$ *is in a Bayesian Nash equilibrium if each player i maximizes her or his expected payoff with respect to her or his belief about the type of the other player given her or his own type:*

$$\psi_i(\lambda_i) \in \underset{\psi \in \Psi_i}{\arg\max} \int_{\lambda_j} f(\lambda|\lambda_i) \cdot \pi_i(\psi, \psi_j(\lambda), \lambda_i, \lambda) \, \mathrm{d}\lambda. \tag{C.3}$$

(see Definition 6.1 in [FT91, p. 215])

For dynamic games, there is a refinement of the Bayesian Nash equilibrium called the *perfect Bayesian equilibrium* which assures the consistent update of beliefs throughout the game to avoid non-credible beliefs and consequently non-credible strategies. The belief's update is based on observed actions of the other player.

Definition C.3 (Perfect Bayesian Equilibrium)
In order for a strategy profile and an associated set of beliefs to be in a perfect Bayesian equilibrium, two requirements have to be met:

- *Sequential rationality of strategies: Each player's strategy has to be determined optimally with respect to the current belief about the other player's type, see Definition C.2.*
- *Consistency of beliefs: The player's belief has to be updated considering observations of the other player's actions.*

(see Definition 8.2 in [FT91, p. 333])

C.2 Additional Lemma on the Sufficient Condition for Maximum Payoff

The following lemma on the sufficient condition for maximum payoff in strategy determination of the applied war of attrition (see Lemma 3.3) is adapted to Fudenberg and Tirole [FT91, pp. 217-218].

Lemma C.1 (Sufficient Condition for Maximum Payoff)
Condition (3.17) is sufficient in terms of maximizing the payoff of (3.16).

Proof:
The sufficiency of condition (3.17) can be proven by contradiction analogous to Fudenberg and Tirole [FT91, pp. 217-218]:

Let $\mathcal{J}_i^*(\tau_i, \delta_i)$ denote the maximum of (3.16). Observe that

$$\frac{\partial^2 \mathcal{J}_i^*(\tau_i, \delta_i)}{\partial \tau_i \partial \delta_i} = f_{\tau_j}(\tau_i) > 0, \quad \forall \tau_i > 0. \tag{C.4}$$

Assume there is another $\tau_i^\diamond$ for which $\mathcal{J}_i^*(\tau_i^\diamond, \delta_i) > \mathcal{J}_i^*(\tau_i, \delta_i)$ holds, given that $\tau_i^\diamond := \tau_i(\delta_i)$. This implies that

$$\int_{\tau_i}^{\tau_i^\diamond} \frac{\partial \mathcal{J}_i^*}{\partial \tau}(\tau, \delta_i)\, \mathrm{d}\tau > 0. \tag{C.5}$$

Together with the first-order condition

$$\frac{\partial \mathcal{J}_i^*}{\partial \tau}(\tau, \phi_i(\tau)) = 0 \quad \forall \tau \tag{C.6}$$

it follows that

$$\int_{\tau_i}^{\tau_i^\diamond} \left(\frac{\partial \mathcal{J}_i^*}{\partial \tau}(\tau, \delta_i) - \frac{\partial \mathcal{J}_i^*}{\partial \tau}(\tau, \phi_i(\tau)) \right) \mathrm{d}\tau > 0$$

and finally that

$$\int_{\tau_i}^{\tau_i^\diamond} \int_{\phi_i(\tau)}^{\delta_i} \frac{\partial^2 \mathcal{J}_i^*(\tau, \delta)}{\partial \tau \partial \delta}\, \mathrm{d}\delta\, \mathrm{d}\tau > 0. \tag{C.7}$$

If $\tau_i^\diamond > \tau_i$ holds, then $\phi_i(\tau) > \delta_i$ follows for all $\tau \in (\tau_i, \tau_i^\diamond]$, which does not fulfill (C.7). This can be derived similarly for $\tau_i^\diamond < \tau_i$. Therefore, τ_i is the global optimum of $\mathcal{J}_i$ for the given utility difference δ_i. □

D Supplements of Cooperative Decision Making Experiments

D.1 Presenting Distributions by Means of Boxplots

By means of a boxplot the distribution of empirical data and related characteristic values can be visualized [Tuk97, pp. 39-43]. Figure D.1 depicts two exemplary compact boxplots of fictional data. A cross denotes the median which divides the dataset

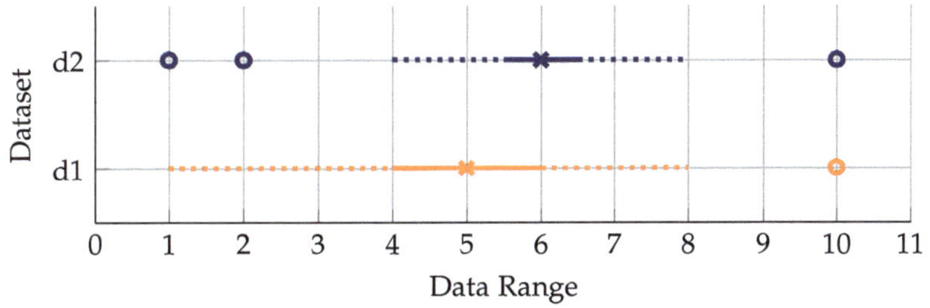

Figure D.1: Exemplary compact boxplots of datasets d1 and d2: Median ×, lower/upper quartile —, lower/upper adjacent · · ·, and outliers ∘.

in half, i. e. 50 % of the data is not smaller or bigger than the median. The box or, in case of the compact boxplot version, a thick line indicate the lower and upper quartiles which form the boundaries of the middle half of the data. This range is called interquartile range. The dots reach out from lower and upper quartile towards lower and upper adjacent, respectively, which are the extreme values of the dataset excluding outliers. Outliers are denoted by circles and are defined as values which have a distance between themselves and the lower or upper quartile that is 1.5-times the length of the box, i. e. the interquartile range.

D.2 Details on the Automation Designs of the Highly Automated Driving Experiment

The following section provides implementation details on the automation designs based on the adaptive negotiation model and the n-stage war of attrition applied in the highly automated driving experiment.

For evaluating the (at most) three possible maneuver options $D \equiv O$ at each intersection with respect to the associated global delays t_{g} and local delays t_{l}, both automation designs applied the following normalized utility function:

$$\bar{u}_i(o) = \bar{u}_i(d) := w_{\mathrm{g},i} \cdot \bar{b}_{\mathrm{g}}(d) + \underbrace{(1 - w_{\mathrm{g},i})}_{w_{\mathrm{l},i}} \cdot \bar{b}_{\mathrm{l}}(d) \tag{D.1a}$$

with
$$\bar{b}_{\mathrm{g}}(d) := \frac{\max_{\forall d^\mu \in D} t_{\mathrm{g}}(d^\mu) - t_{\mathrm{g}}(d)}{\max_{\forall d^\mu \in D} t_{\mathrm{g}}(d^\mu) - \min_{\forall d^\mu \in D} t_{\mathrm{g}}(d^\mu)}, \tag{D.1b}$$

$$\bar{b}_{\mathrm{l}}(d) := \frac{\max_{\forall d^\mu \in D} t_{\mathrm{l}}(d^\mu) - t_{\mathrm{l}}(d)}{\max_{\forall d^\mu \in D} t_{\mathrm{l}}(d^\mu) - \min_{\forall d^\mu \in D} t_{\mathrm{l}}(d^\mu)}. \tag{D.1c}$$

Based on this utility evaluation, the utility difference distribution $f_{\delta_\mathcal{A}}$ for the automation design based on the n-stage war of attrition was determined by aggregating all utility differences $\delta_\mathcal{A}$ of all intersections of the Manhattan grid. The utility difference distribution $f_{\delta_\mathcal{H}}$ was initially set to a uniform distribution within the range of value of $f_{\delta_\mathcal{A}}$ and was subsequently updated analogous to the identification algorithm described in Section 4.2.3. Furthermore, on the basis of the results of the suitability study (see Sections 4.1.3 and 4.2.3) in terms of exponential cost function fit, the cost function was set to be quadratic, i. e.

$$c(t) \sim t^2. \tag{D.2}$$

The prefactor of the cost function was determined for each decision scenario according to the procedure described in Section 4.2.3.

As the decision making process was set to start when the human initially chose a maneuver option at time t_0, the time normalization required for the target utility (3.3) of the adaptive negotiation model as well as for the cost function (D.2) was defined as follows:

$$\bar{t} = \frac{t}{\mathcal{T} - t_0}. \tag{D.3}$$

The parameters of the automation design based on the adaptive negotiation model introduced in Sections 3.2 and 4.2.2 were partially inspired by the results of the suitability study (see Sections 4.1.2 and 4.2.2) and are summarized in Table D.1. For the identification of the human behavior, the same utility function structure and target utility structure as for the automation design based on the adaptive negotiation model were assumed.

D.3 Questionnaires of the Highly Automated Driving Experiment

The translated questionnaires for the highly automated driving experiment are depicted in Figures D.2 to D.5: the first questionnaire is concerned with general and

Table D.1: Parameters of the adaptive negotiation model in the highly automated driving experiment.

Parameter Description	**Value/Range**
Utility weight $w_{g,\mathcal{A}}$	1
Initial concession rate $\epsilon_{\mathcal{A}}$	0.3
Estimation range for utility weight $\hat{w}_{g,\mathcal{H}}$	$[0, 0.2, \ldots, 1]$
Estimation range of concession rate $\hat{\epsilon}_{\mathcal{H}}$	$[0.1, 0.3, \ldots, 1.1]$
Adaptation parameter $\beta_{\mathcal{A}}$	0.5
Risk disposition factor $r_{\mathcal{A}}$	0.2
Adaptation range of concession rate $\epsilon_{\mathcal{A}}$	$[0.1, 0.2, \ldots, 0.5]$

personal information, the second and third are filled out after each experimental run and the fourth questionnaire is for comparing all experimental runs.

KIT
Karlsruhe Institute of Technology

IRS

Questionnaire of the Cooperative Decision Making Experiment

Please check all that apply and elaborate where needed.

Before the Start of the Experiment

- Age: _______
- Gender: □ male □ female □ diverse
- Profession: ____
- Possession of a valid driver license: □ Yes □ No
- Experience with driving simulators: □ Yes □ No
  If yes, which? ____
- Do you have experience with navigation systems? □ Yes □ No
- How do you assess your style of driving?

Fast	□	□	□	□	□	Slow
Proactive	□	□	□	□	□	Reactive
Aggressive	□	□	□	□	□	Defensive

- How do you judge your cognitive capabilities to assess traffic situations?
  For example the correct assessment of danger and deceleration of other road users.

Good	□	□	□	□	□	Bad

- You have to make a decision in a group or team.
  How do you judge your demeanor in pushing through your interests?

Dominant	□	□	□	□	□	Restrained

  How do you assess your willingness to compromise in such situations?

High	□	□	□	□	□	Low

We wish you a pleasant time while participating in the experiment!

KIT – The Research University in the Helmholtz Association www.kit.edu

Figure D.2: Questionnaire for general and personal information.

After Each Experimental Run

- Did you follow the instructions? *Did you make an effort to find the shortest path to the destination by means of cooperating with your cooperation partner?*
  □ Yes □ Partially □ No

- You have agreed multiple times with your partner on a decision option.
  How do you asses **yourself** in the cooperative decision making?
  Ready to compromise □ □ □ □ □ Stubborn

  How do you judge **your partner** in the cooperative decision?
  Ready to compromise □ □ □ □ □ Stubborn

  How do you assess the overall cooperation between you and your partner?
  Satisfactory □ □ □ □ □ Dissatisfactory

- Did you follow a specific strategy while making cooperative decisions?
  □ Yes □ No
  If yes, which one? ______________________

- Did perceive the cooperative decision making in this experimental run as negative, neutral or positive with respect of reaching the destination in the shortest time possible?
  □ Positive □ Neutral □ Negative

- Was your behavior in cooperative decision making transparent?
  Transparent □ □ □ □ □ Non-transparent

- Was the behavior of your partner in cooperative decision making transparent?
  Transparent □ □ □ □ □ Non-transparent

- How do you assess your partner's cooperation behavior?
  Reliable □ □ □ □ □ Not reliable

- How did you assess the interaction between you and your partner?
  Intuitive □ □ □ □ □ Not intuitive

  If not or little intuitive, why? ______________________

- How did you perceive the recent experimental run?
  Very strenuous □ □ □ □ □ Very easy

KIT – The Research University in the Helmholtz Association www.kit.edu

Figure D.3: Questionnaire after each experimental run: first page.

- Did you observe any change in your partner's behavior during the last experimental run?
  □ Yes □ No
  If yes, how do you assess this change?
  Positive □ □ □ □ □ Negative

- Please leave some notes for the later comparison of the experimental runs (e.g. adjectives describing the last experimental run).

  ____

  ____

  ____

KIT – The Research University in the Helmholtz Association www.kit.edu

Figure D.4: Questionnaire after each experimental run: second page.

After Finishing All Experimental Runs

Please answer the following question again but with focus on the relation between the four experimental runs (ER1 – ER4).

- How do you assess the overall cooperation between you and your partner?

Satisfactory	ER1	□	□	□	□	□	Dissatisfactory
	ER2	□	□	□	□	□	
	ER3	□	□	□	□	□	
	ER4	□	□	□	□	□	

- Was the behavior of your partner in cooperative decision making transparent?

Transparent	ER1	□	□	□	□	□	Non transparent
	ER2	□	□	□	□	□	
	ER3	□	□	□	□	□	
	ER4	□	□	□	□	□	

- How do you assess your partner's cooperation behavior?

Reliable	ER1	□	□	□	□	□	Not reliable
	ER2	□	□	□	□	□	
	ER3	□	□	□	□	□	
	ER4	□	□	□	□	□	

- How did you assess the interaction between you and your partner?

Intuitive	ER1	□	□	□	□	□	Not intuitive
	ER2	□	□	□	□	□	
	ER3	□	□	□	□	□	
	ER4	□	□	□	□	□	

- How did you assess the experimental runs in comparison?

Very strenuous	ER1	□	□	□	□	□	Very easy
	ER2	□	□	□	□	□	
	ER3	□	□	□	□	□	
	ER4	□	□	□	□	□	

The Institute of Control Systems thanks you for your participations!

KIT – The Research University in the Helmholtz Association www.kit.edu

Figure D.5: Questionnaire for comparison of all experimental runs.

References

Public References

[ABC+16] ALTENDORF, E.; BALTZER, M.; CANPOLAT, Y.; LOPEZ, D.; SCHRECK, C.; WESSEL, G.; FLEMISCH, F.: Automated Driving using Shared and Cooperative Guidance and Control: Revisiting past Research with a new Model. In: *IFAC-PapersOnLine* 49 (2016), Nr. 19, S. 165–170. http://dx.doi.org/10.1016/j.ifacol.2016.10.480. – DOI 10.1016/j.ifacol.2016.10.480. – ISSN 24058963

[ABH+16] ALTENDORF, E.; BALTZER, M.; HEESEN, M.; KIENLE, M.; WEISSGERBER, T.; FLEMISCH, F.: H-Mode. Version: 2016. http://dx.doi.org/10.1007/978-3-319-12352-3{_}60. In: WINNER, H. (Hrsg.); HAKULI, S. (Hrsg.); LOTZ, F. (Hrsg.); SINGER, C. (Hrsg.): *Handbook of Driver Assistance Systems*. Cham: Springer International Publishing, 2016. – DOI 10.1007/978–3–319–12352–3_60. – ISBN 978–3–319–12352–3, 1499–1518

[ACM+18] ABBINK, D. A.; CARLSON, T.; MULDER, M.; WINTER, J. C. F.; AMINRAVAN, F.; GIBO, T. L.; BOER, E. R.: A Topology of Shared Control Systems—Finding Common Ground in Diversity. In: *IEEE Transactions on Human-Machine Systems* 48 (2018), Nr. 5, S. 509–525. http://dx.doi.org/10.1109/THMS.2018.2791570. – DOI 10.1109/THMS.2018.2791570. – ISSN 2168–2291

[AF80] AJZEN, I.; FISHBEIN, M.: *Understanding attitudes and predicting social behavior*. Englewood Cliffs, NJ: Prentice-Hall, 1980. – ISBN 0–13–936435–8

[AG00] ABREU, D.; GUL, F.: Bargaining and Reputation. In: *Econometrica* 68 (2000), Nr. 1, S. 85–117. http://dx.doi.org/10.1111/1468-0262.00094. – DOI 10.1111/1468–0262.00094. – ISSN 00129682

[AGH04] ANDERSON, S. P.; GOEREE, J. K.; HOLT, C. A.: Noisy Directional Learning and the Logit Equilibrium. In: *The Scandinavian Journal of Economics* 106 (2004), Nr. 3, S. 581–602. http://dx.doi.org/10.1111/j.0347-0520.2004.00378.x. – DOI 10.1111/j.0347–0520.2004.00378.x. – ISSN 0347–0520

[AH81] AXELROD, R.; HAMILTON, W.: The evolution of cooperation. In: *Science* 211 (1981), Nr. 4489, S. 1390–1396. http://dx.doi.org/10.1126/science.7466396. – DOI 10.1126/science.7466396. – ISSN 0036–8075

[Ajz85] AJZEN, I.: From Intentions to Actions: A Theory of Planned Behavior. Version: 1985. http://dx.doi.org/10.1007/978-3-642-69746-3{_}2. In: KUHL, J. (Hrsg.); BECKMANN, J. (Hrsg.): *Action Control* Bd. 34. Berlin, Heidelberg: Springer Berlin Heidelberg, 1985. – DOI 10.1007/978–3–642–69746–3_2. – ISBN 978–3–642–69748–7, S. 11–39

[AM06] AGASTYA, M. K.; MCAFEE, R. P.: Continuing Wars of Attrition. In: *SSRN Electronic Journal* (2006). http://dx.doi.org/10.2139/ssrn.935970. – DOI 10.2139/ssrn.935970

[AMB12] ABBINK, D. A.; MULDER, M.; BOER, E. R.: Haptic shared control: Smoothly shifting control authority? In: *Cognition, Technology & Work* 14 (2012), Nr. 1, S. 19–28. http://dx.doi.org/10.1007/s10111-011-0192-5. – DOI 10.1007/s10111–011–0192–5. – ISSN 1435–5558

[ASM+05] ADLER, J. L.; SATAPATHY, G.; MANIKONDA, V.; BOWLES, B.; BLUE, V. J.: A multi-agent approach to cooperative traffic management and route guidance. In: *Transportation Research Part B: Methodological* 39 (2005), Nr. 4, S. 297–318. http://dx.doi.org/10.1016/j.trb.2004.03.005. – DOI 10.1016/j.trb.2004.03.005. – ISSN 01912615

[AY21] ALBABA, B. M.; YILDIZ, Y.: Driver Modeling Through Deep Reinforcement Learning and Behavioral Game Theory. In: *IEEE Transactions on Control Systems Technology* (2021), S. 1–8. http://dx.doi.org/10.1109/TCST.2021.3075557. – DOI 10.1109/TCST.2021.3075557. – ISSN 1063–6536

[Baa16] BAARSLAG, T.: *Exploring the Strategy Space of Negotiating Agents.* Cham: Springer International Publishing, 2016. http://dx.doi.org/10.1007/978-3-319-28243-5. http://dx.doi.org/10.1007/978-3-319-28243-5. – ISBN 978–3–319–28242–8

[Bai82] BAINBRIDGE, L.: Ironies of automation. In: *IFAC Proceedings Volumes* 15 (1982), Nr. 6, S. 129–135. – ISSN 14746670

[BAMF14] BALTZER, M.; ALTENDORF, E. E. O.; MEIER, S.; FLEMISCH, F.: Mediating the Interaction between Human and Automation during the Arbitration Processes in Cooperative Guidance and Control of Highly Automated Vehicles: Base concept and First Study. Version: 2014. http://publications.rwth-aachen.de/record/459815. In: *Proceedings of the 5th International Conference on Applied Human Factors and*

Ergonomics : AHFE 2014 ; 19-23 July, Kraków, Poland / T. Ahram, W. Karwowski and T. Marek. AHFE, 2014, 2107–2118

[BC78] Bishop, D. T.; Cannings, C.: A generalized war of attrition. In: *Journal of Theoretical Biology* 70 (1978), Nr. 1, S. 85–124. http://dx.doi.org/10.1016/0022-5193(78)90304-1. – DOI 10.1016/0022–5193(78)90304–1. – ISSN 00225193

[BCM78] Bishop, D. T.; Cannings, C.; Maynard Smith, J.: The war of attrition with random rewards. In: *Journal of Theoretical Biology* 74 (1978), Nr. 3, S. 377–388. http://dx.doi.org/10.1016/0022-5193(78)90220-5. – DOI 10.1016/0022–5193(78)90220–5. – ISSN 00225193

[BD16] Bruder, R.; Didier, M.: Design of Human-Machine-Interfaces for DAS. In: Winner, H. (Hrsg.); Hakuli, S. (Hrsg.); Lotz, F. (Hrsg.); Singer, C. (Hrsg.): *Handbook of Driver Assistance Systems*. Cham: Springer International Publishing, 2016. – ISBN 978–3–319–12352–3, S. 797–812

[BFH19] Braun, C. A.; Flad, M.; Hohmann, S.: A Continuous and Quantitative Metric for the Levels of Automation. In: *IFAC-PapersOnLine* 52 (2019), Nr. 19, S. 37–42. http://dx.doi.org/10.1016/j.ifacol.2019.12.081. – DOI 10.1016/j.ifacol.2019.12.081. – ISSN 24058963

[Bil96] Billings, C. E.: *Human-Centered Aviation Automation: Principles and Guidelines*. 1996 https://ntrs.nasa.gov/citations/19960016374

[BK99] Bulow, J.; Klemperer, P.: The Generalized War of Attrition. In: *The American Economic Review* 89 (1999), Nr. 1, 175–189. http://www.jstor.org/stable/116984. – ISSN 00028282

[BK17] Bütepage, J.; Kragic, D.: *Human-Robot Collaboration: From Psychology to Social Robotics*. https://arxiv.org/abs/1705.10146. Version: 2017

[BLF19] Baltzer, M. C. A.; López, D.; Flemisch, F.: Towards an interaction pattern language for human machine cooperation and cooperative movement. In: *Cognition, Technology & Work* 21 (2019), Nr. 4, S. 593–606. http://dx.doi.org/10.1007/s10111-019-00561-8. – DOI 10.1007/s10111–019–00561–8. – ISSN 1435–5566

[BLP16] Bhattacharya, R. N.; Lin, L.; Patrangenaru, V.: *A Course in Mathematical Statistics and Large Sample Theory*. New York: Springer, 2016 (Springer Texts in Statistics). http://dx.doi.org/10.1007/978-1-4939-4032-5. http://dx.doi.org/10.1007/978-1-4939-4032-5. – ISBN 978–1–4939–4032–5

[Bro86] Brooks, R.: A robust layered control system for a mobile robot. In: *IEEE Journal on Robotics and Automation* 2 (1986), Nr. 1, S. 14–23. http://dx.doi.org/10.1109/JRA.1986.1087032. – DOI 10.1109/JRA.1986.1087032. – ISSN 0882–4967

[BSMM15] Bronshtein, I. N.; Semendyayev, K. A.; Musiol, G.; Mühlig, H.: *Handbook of Mathematics*. Berlin, Heidelberg: Springer Berlin Heidelberg, 2015. http://dx.doi.org/10.1007/978-3-662-46221-8. http://dx.doi.org/10.1007/978-3-662-46221-8. – ISBN 978–3–662–46220–1

[Bub03] Bubb, H.: Ergonomics and human reliability. In: *Proceedings of the Institution of Mechanical Engineers, Part B: Journal of Engineering Manufacture* 217 (2003), Nr. 11, S. 1627–1632. http://dx.doi.org/10.1243/095440503771910005. – DOI 10.1243/095440503771910005. – ISSN 0954–4054

[BvD+14] Bloem, J.; van Doorn, M.; Duivestein, S.; Excoffier, D.; Maas, R.; van Ommeren, E.: The fourth industrial revolution. In: *Things Tighten* 8 (2014), S. 11–15

[BYK+02] Burghart, C.; Yigit, S.; Kerpa, O.; Osswald, D.; Woern, H.: Concept for human robot co-operation integrating artificial haptic perception. In: *Intelligent Autonomous Systems* Bd. 7, 2002, S. 38–45

[CATW13] Chen, S.; Ammar, H. B.; Tuyls, K.; Weiss, G.: Optimizing Complex Automated Negotiation Using Sparse Pseudo-input Gaussian Processes. In: *Proceedings of the 2013 International Conference on Autonomous Agents and Multi-agent Systems*. Richland, SC: International Foundation for Autonomous Agents and Multiagent Systems, 2013 (AAMAS '13). – ISBN 978–1–4503–1993–5, 707–714

[CD13] Carlson, T.; Del R. Millan, J.: Brain-Controlled Wheelchairs: A Robotic Architecture. In: *IEEE Robotics & Automation Magazine* 20 (2013), Nr. 1, S. 65–73. http://dx.doi.org/10.1109/MRA.2012.2229936. – DOI 10.1109/MRA.2012.2229936. – ISSN 1070–9932

[CE08] Caruana, G.; Einav, L.: A Theory of Endogenous Commitment. In: *The Review of Economic Studies* 75 (2008), Nr. 1, 99–116. http://www.jstor.org/stable/4626189. – ISSN 00346527

[CGC06] Costa-Gomes, M. A.; Crawford, V. P.: Cognition and Behavior in Two-Person Guessing Games: An Experimental Study. In: *The American Economic Review* 96 (2006), Nr. 5, S. 1737–1768. http://dx.doi.org/10.1257/aer.96.5.1737. – DOI 10.1257/aer.96.5.1737. – ISSN 00028282

[CGIC09] Costa-Gomes, M. A.; Iriberri, N.; Crawford, V. P.: Comparing Models of Strategic Thinking in Van Huyck, Battalio, and Beil's Coordination Games. In: *Journal of the European Economic Association* 7 (2009), Nr. 2/3, 365–376. http://www.jstor.org/stable/40282755. – ISSN 15424766

[CHC04] Camerer, C. F.; Ho, T.-H.; Chong, J.-K.: A Cognitive Hierarchy Model of Games. In: *The Quarterly Journal of Economics* 119 (2004), Nr. 3, S. 861–898. http://dx.doi.org/10.1162/0033553041502225. – DOI 10.1162/0033553041502225. – ISSN 0033–5533

[CHC16] Chong, J.-K.; Ho, T.-H.; Camerer, C.: A generalized cognitive hierarchy model of games. In: *Games and Economic Behavior* 99 (2016), S. 257–274. http://dx.doi.org/10.1016/j.geb.2016.08.007. – DOI 10.1016/j.geb.2016.08.007. – ISSN 08998256

[CHS21] Chiou, M.; Hawes, N.; Stolkin, R.: Mixed-initiative Variable Autonomy for Remotely Operated Mobile Robots. In: *ACM transactions on human-robot interaction* 10 (2021), Nr. 4, S. 1–34. http://dx.doi.org/10.1145/3472206. – DOI 10.1145/3472206

[CJ04] Coehoorn, R. M.; Jennings, N. R.: Learning on opponent's preferences to make effective multi-issue negotiation trade-offs. In: Janssen, M. (Hrsg.); Sol, H. G. (Hrsg.); Wagenaar, R. W. (Hrsg.): *Proceedings of the 6th international conference on Electronic commerce - ICEC '04*. New York, New York, USA: ACM Press, 2004. – ISBN 1581139306, S. 59

[CKLS14] Calcagno, R.; Kamada, Y.; Lovo, S.; Sugaya, T.: Asynchronicity and coordination in common and opposing interest games. In: *Theoretical Economics* 9 (2014), Nr. 2, S. 409–434. http://dx.doi.org/10.3982/TE1202. – DOI 10.3982/TE1202. – ISSN 19336837

[Cow01] Cowan, N.: The magical number 4 in short-term memory: a reconsideration of mental storage capacity. In: *The Behavioral and brain sciences* 24 (2001), Nr. 1, S. 87–114; discussion 114–85. http://dx.doi.org/10.1017/S0140525X01003922. – DOI 10.1017/S0140525X01003922. – ISSN 0140–525X

[CPRMML17] Cruz-Piris, L.; Rivera, D.; Marsá-Maestre, I.; La Hoz, E. de: Optimizing vehicle trips using agent negotiation through a traffic matrix. In: *The Tenth International Workshop on Agent-based Complex Automated Negotiations (ACAN2017)*. São Paulo (Brazil), 2017

[CRDD20] Chanel, C. P. C.; Roy, R. N.; Drougard, N.; Dehais, F.: Mixed-Initiative Human-Automated Agents Teaming: Towards a Flexible Cooperation Framework. Version: 2020. http://dx.doi.org/10.

1007/978-3-030-49183-3{_}10. In: HARRIS, D. (Hrsg.); LI, W.-C. (Hrsg.): *Engineering Psychology and Cognitive Ergonomics. Cognition and Design* Bd. 12187. Cham: Springer International Publishing, 2020. – DOI 10.1007/978-3-030-49183-3_10. – ISBN 978-3-030-49182-6, S. 117–133

[CRN12] CHATTERJEE, K.; REITER, J. G.; NOWAK, M. A.: Evolutionary dynamics of biological auctions. In: *Theoretical population biology* 81 (2012), Nr. 1, S. 69–80. http://dx.doi.org/10.1016/j.tpb.2011.11.003. – DOI 10.1016/j.tpb.2011.11.003

[CSB+16] CHIOU, M.; STOLKIN, R.; BIEKSAITE, G.; HAWES, N.; SHAPIRO, K. L.; HARRISON, T. S.: Experimental analysis of a variable autonomy framework for controlling a remotely operating mobile robot. In: *2016 IEEE/RSJ International Conference on Intelligent Robots and Systems (IROS)*, IEEE, 2016. – ISBN 978-1-5090-3762-9, S. 3581–3588

[CSP14] CORREDOR, J.; SOFRONY, J.; PEER, A.: Deciding on optimal assistance policies in haptic shared control tasks. In: *2014 IEEE International Conference on Robotics and Automation (ICRA)*, IEEE, 2014. – ISBN 978-1-4799-3685-4, S. 2679–2684

[CTS19] CHIOU, M.; TALHA, M.; STOLKINL, R.: Learning effects in variable autonomy human-robot systems: how much training is enough? In: *2019 IEEE International Conference on Systems, Man and Cybernetics (SMC)*, IEEE, 2019. – ISBN 978-1-7281-4569-3, S. 720–727

[CW15] CHEN, S.; WEISS, G.: An approach to complex agent-based negotiations via effectively modeling unknown opponents. In: *Expert Systems with Applications* 42 (2015), Nr. 5, S. 2287–2304. http://dx.doi.org/10.1016/j.eswa.2014.10.048. – DOI 10.1016/j.eswa.2014.10.048. – ISSN 09574174

[DLDS13] DESAI, P.; LOKE, S. W.; DESAI, A.; SINGH, J.: CARAVAN: Congestion Avoidance and Route Allocation Using Virtual Agent Negotiation. In: *IEEE Transactions on Intelligent Transportation Systems* 14 (2013), Nr. 3, S. 1197–1207. http://dx.doi.org/10.1109/TITS.2013.2256420. – DOI 10.1109/TITS.2013.2256420. – ISSN 1524-9050

[DM94] DAVID, P.; MONROE, H.: Standards development strategies under incomplete information. In: *MERIT Research Memorandum*. University of Limburg Maastricht, 1994

[Don82] DONGES, E.: Aspekte der aktiven Sicherheit bei der Führung von Personenkraftwagen. In: *Automob-Ind* 27 (1982), Nr. 2

[Don99] Donges, E.: A Conceptual Framework for Active Safety in Road Traffic. In: *Vehicle System Dynamics* 32 (1999), Nr. 2-3, S. 113–128. `http://dx.doi.org/10.1076/vesd.32.2.113.2089`. – DOI 10.1076/vesd.32.2.113.2089. – ISSN 0042–3114

[DP14] During, M.; Pascheka, P.: Cooperative decentralized decision making for conflict resolution among autonomous agents. In: *2014 IEEE International Symposium on Innovations in Intelligent Systems and Applications (INISTA) Proceedings*, IEEE, 2014. – ISBN 978–1–4799–3020–3, S. 154–161

[DS13] Dragan, A. D.; Srinivasa, S. S.: A policy-blending formalism for shared control. In: *The International Journal of Robotics Research* 32 (2013), Nr. 7, S. 790–805. `http://dx.doi.org/10.1177/0278364913490324`. – DOI 10.1177/0278364913490324. – ISSN 0278–3649

[DvA+10] Della Penna, M.; van Paassen, M. M.; Abbink, D. A.; Mulder, M.; Mulder, M.: Reducing steering wheel stiffness is beneficial in supporting evasive maneuvers. In: *2010 IEEE International Conference on Systems, Man and Cybernetics*, IEEE, 2010. – ISBN 978–1–4244–6586–6, S. 1628–1635

[EK99] Endsley, M. R.; Kaber, D. B.: Level of automation effects on performance, situation awareness and workload in a dynamic control task. In: *Ergonomics* 42 (1999), Nr. 3, S. 462–492. `http://dx.doi.org/10.1080/001401399185595`. – DOI 10.1080/001401399185595

[End87] Endsley, M. R.: The Application of Human Factors to the Development of Expert Systems for Advanced Cockpits. In: *Proceedings of the Human Factors Society Annual Meeting* 31 (1987), Nr. 12, S. 1388–1392. `http://dx.doi.org/10.1177/154193128703101219`. – DOI 10.1177/154193128703101219. – ISSN 0163–5182

[End17] Endsley, M. R.: From Here to Autonomy. In: *Human factors* 59 (2017), Nr. 1, S. 5–27. `http://dx.doi.org/10.1177/0018720816681350`. – DOI 10.1177/0018720816681350. – ISSN 0018–7208

[FA75] Fishbein, M.; Ajzen, I.: *Belief, attitude, intention and behaviour: An introduction to theory and research*. Bd. 27. 1975

[FAC+03] Flemisch, F. O.; Adams, C. A.; Conway, S. R.; Goodrich, K. H.; Palmer, M. T.; Schutte, P. C.: The H-Metaphor as a guideline for vehicle automation and interaction. (2003)

[FAI+16] Flemisch, F.; Abbink, D.; Itoh, M.; Pacaux-Lemoine, M.-P.; Wessel, G.: Shared control is the sharp end of cooperation: Towards a common framework of joint action, shared control and human machine cooperation. In: *IFAC-PapersOnLine* 49 (2016), Nr. 19, S. 72–77. http://dx.doi.org/10.1016/j.ifacol.2016.10.464. – DOI 10.1016/j.ifacol.2016.10.464. – ISSN 24058963

[FAI+19] FLEMISCH, F.; ABBINK, D. A.; ITOH, M.; PACAUX-LEMOINE, M.-P.; WESSEL, G.: Joining the blunt and the pointy end of the spear: towards a common framework of joint action, human–machine cooperation, cooperative guidance and control, shared, traded and supervisory control. In: *Cognition, Technology & Work* 21 (2019), Nr. 4, S. 555–568. http://dx.doi.org/10.1007/s10111-019-00576-1. – DOI 10.1007/s10111–019–00576–1. – ISSN 1435–5566

[FBB+14] FLEMISCH, F. O.; BENGLER, K.; BUBB, H.; WINNER, H.; BRUDER, R.: Towards cooperative guidance and control of highly automated vehicles: H-Mode and Conduct-by-Wire. In: *Ergonomics* 57 (2014), Nr. 3, S. 343–360. http://dx.doi.org/10.1080/00140139.2013.869355. – DOI 10.1080/00140139.2013.869355

[FCA+17] FLEMISCH, F.; CANPOLAT, Y.; ALTENDORF, E.; WESEL, G.; ITOH, M.; BALTZER, M.; PACAUX-LEMOINE, M.-P.; ABBINK, D.; SCHUTTE, P.: Shared and cooperative control of ground and air vehicles: Introduction and general overview. In: *2017 IEEE International Conference on Systems, Man, and Cybernetics (SMC)*, IEEE, 2017. – ISBN 978–1–5386–1645–1, S. 858–863

[FDM+20] FLEMISCH, F.; DIEDERICHS, F.; MEYER, R.; HERZBERGER, N.; BAIER, R.; ALTENDORF, E.; SPIES, J.; USAI, M.; KAIM, V.; DOENTGEN, B.; BOPP-BERTENBREITER, A. V.; WIDLROITHER, H.; RUTH-SCHUMACHER, S.; ARZT, C.; BOZBAYIR, E.; BISCHOFF, S.; DIERS, D.; WECHNER, R.; SOMMER, A.; AYDIN, E.; KASCHUB, V.; KIEFER, T.; HOTTELART, K.; REILHAC, P.; WESSEL, G.; KAISER, F.: Vorreiter: Manoeuvre-Based Steering Gestures for Partially and Highly Automated Driving. Version: 2020. http://dx.doi.org/10.1007/978-3-030-45131-8{_}5. In: MEIXNER, G. (Hrsg.): *Smart Automotive Mobility*. Cham: Springer International Publishing, 2020 (Human–Computer Interaction Series). – DOI 10.1007/978–3–030–45131–8_5. – ISBN 978–3–030–45130–1, S. 231–304

[Fin04] FINK, A.: Supply chain coordination by means of automated negotiations. In: *37th Annual Hawaii International Conference on System Sciences, 2004. Proceedings of the*, IEEE, 2004. – ISBN 0–7695–2056–1, S. 10 pp

[FIZ+16] FUKUTA, N.; ITO, T.; ZHANG, M.; FUJITA, K.; ROBU, V.: *Recent Advances in Agent-based Complex Automated Negotiation*. Bd. 638.

Cham: Springer International Publishing, 2016. http://dx.doi.org/10.1007/978-3-319-30307-9. http://dx.doi.org/10.1007/978-3-319-30307-9. – ISBN 978–3–319–30305–5

[FKB+12] Franz, B.; Kauer, M.; Blanke, A.; Schreiber, M.; Bruder, R.; Geyer, S.: Comparison of two human-machine-interfaces for cooperative maneuver-based driving. In: *Work (Reading, Mass.)* 41 Suppl 1 (2012), S. 4192–4199. http://dx.doi.org/10.3233/WOR-2012-0121-4192. – DOI 10.3233/WOR–2012–0121–4192

[FKBG12] Franz, B.; Kauer, M.; Bruder, R.; Geyer, S.: pieDrive - a New Driver-Vehicle Interaction Concept for Maneuver-Based Driving. In: *2012 International IEEE Intelligent Vehicles Symposium Workshops*. Alcalá de Henares, Spanien, 2012

[FKGH15] Franz, B.; Kauer, M.; Geyer, S.; Hakuli, S.: Conduct-by-Wire. Version: 2015. http://dx.doi.org/10.1007/978-3-658-05734-3{_}59. In: Winner, H. (Hrsg.); Hakuli, S. (Hrsg.); Lotz, F. (Hrsg.); Singer, C. (Hrsg.): *Handbuch Fahrerassistenzsysteme: Grundlagen, Komponenten und Systeme für aktive Sicherheit und Komfort*. Wiesbaden: Springer Fachmedien Wiesbaden, 2015. – DOI 10.1007/978–3–658–05734–3_59. – ISBN 978–3–658–05734–3, S. 1111–1121

[Fla17] Flad, M.: *Kooperative Regelungskonzepte auf Basis der Spieltheorie und deren Anwendung auf Fahrerassistenzsysteme*. Karlsruhe, Karlsruher Institut für Technologie, Diss., 2017. http://dx.doi.org/10.5445/KSP/1000063300. – DOI 10.5445/KSP/1000063300

[Fla19] Flad, M.: Differential-Game-Based Driver Assistance System for Fuel-Optimal Driving. Version: 2019. http://dx.doi.org/10.1007/978-3-030-23699-1{_}2. In: Petrosyan, L. A. (Hrsg.); Mazalov, V. V. (Hrsg.); Zenkevich, N. A. (Hrsg.): *Frontiers of Dynamic Games*. Cham: Springer International Publishing, 2019 (Static & Dynamic Game Theory: Foundations & Applications). – DOI 10.1007/978–3–030–23699–1_2. – ISBN 978–3–030–23698–4, S. 13–36

[FNJT07] Fern, A.; Natarajan, S.; Judah, K.; Tadepalli, P.: A Decision-Theoretic Model of Assistance. In: *Proceedings of the 20th International Joint Conference on Artifical Intelligence*. San Francisco, CA, USA: Morgan Kaufmann Publishers Inc, 2007 (IJCAI'07), S. 1879–1884

[FPLV+20] Flemisch, F. O.; Pacaux-Lemoine, M.-P.; Vanderhaegen, F.; Itoh, M.; Saito, Y.; Herzberger, N.; Wasser, J.; Grislin, E.; Baltzer, M.: Conflicts in Human-Machine Systems as an Intersection of Bio- and Technosphere: Cooperation and Interaction Patterns for Human and

Machine Interference and Conflict Resolution. In: *2020 IEEE International Conference on Human-Machine Systems (ICHMS)*, IEEE, 2020. – ISBN 978–1–7281–5871–6, S. 1–6

[Fri11] Friston, K.: What is optimal about motor control? In: *Neuron* 72 (2011), Nr. 3, S. 488–498. http://dx.doi.org/10.1016/j.neuron.2011.10.018. – DOI 10.1016/j.neuron.2011.10.018. – ISSN 0896–6273

[FS48] Friedman, M.; Savage, L. J.: The Utility Analysis of Choices Involving Risk. In: *Journal of Political Economy* 56 (1948), Nr. 4, 279–304. http://www.jstor.org/stable/1826045. – ISSN 00223808

[FSJ98] Faratin, P.; Sierra, C.; Jennings, N. R.: Negotiation decision functions for autonomous agents. In: *Robotics and Autonomous Systems* 24 (1998), Nr. 3-4, S. 159–182. http://dx.doi.org/10.1016/S0921-8890(98)00029-3. – DOI 10.1016/S0921–8890(98)00029–3. – ISSN 09218890

[FSKL08] Flemisch, F.; Schieben, A.; Kelsch, J.; Löper, C.: Automation spectrum, inner/outer compatibility and other potentially useful human factors concepts for assistance and automation. In: *Human Factors for assistance and automation* (2008)

[FT86] Fudenberg, D.; Tirole, J.: A Theory of Exit in Duopoly. In: *Econometrica* 54 (1986), Nr. 4, 943–960. http://dx.doi.org/10.2307/1912845. – DOI 10.2307/1912845. – ISSN 00129682

[FT91] Fudenberg, D.; Tirole, J.: *Game theory*. Cambridge, Mass.: MIT Press, 1991. – ISBN 9780262061414

[FWBB16] Flemisch, F.; Winner, H.; Bruder, R.; Bengler, K.: Cooperative guidance, control and automation. In: Winner, H. (Hrsg.); Hakuli, S. (Hrsg.); Lotz, F. (Hrsg.); Singer, C. (Hrsg.): *Handbook of Driver Assistance Systems*. Cham: Springer International Publishing, 2016. – ISBN 978–3–319–12352–3, S. 1471–1481

[GAB15] Gaciarz, M.; Aknine, S.; Bhouri, N.: Constraint-Based Negotiation Model for Traffic Regulation. In: *2015 IEEE/WIC/ACM International Conference on Web Intelligence and Intelligent Agent Technology (WI-IAT)*, IEEE, 2015. – ISBN 978–1–4673–9618–9, S. 320–327

[GAR86] Greenstein, J. S.; Arnaut, L. Y.; Revesman, M. E.: An empirical comparison of model-based and explicit communication for dynamic human-computer task allocation. In: *International Journal of Man-Machine Studies* 24 (1986), Nr. 4, S. 355–363. http://dx.doi.org/10.1016/S0020-7373(86)80051-7. – DOI 10.1016/S0020–7373(86)80051–7. – ISSN 00207373

[GCWF18] Gallivan, J. P.; Chapman, C. S.; Wolpert, D. M.; Flanagan, J. R.: Decision-making in sensorimotor control. In: *Nature reviews. Neuroscience* 19 (2018), Nr. 9, S. 519–534. `http://dx.doi.org/10.1038/s41583-018-0045-9`. – DOI 10.1038/s41583–018–0045–9

[GDLB13] Gold, C.; Damböck, D.; Lorenz, L.; Bengler, K.: "Take over!" How long does it take to get the driver back into the loop? In: *Proceedings of the Human Factors and Ergonomics Society Annual Meeting* 57 (2013), Nr. 1, S. 1938–1942. `http://dx.doi.org/10.1177/1541931213571433`. – DOI 10.1177/1541931213571433. – ISSN 1541–9312

[GDW91] Gmytrasiewicz, P. J.; Durfee, E. H.; Wehe, D. K.: Combining decision theory and hierarchical planning for a time-dependent robotic application. In: *The Seventh IEEE Conference on Artificial Intelligence Application Proceedings*, IEEE Comput. Soc. Press, 1991. – ISBN 0–8186–2135–4, S. 282–288

[GFKP13] Groten, R.; Feth, D.; Klatzky, R. L.; Peer, A.: The role of haptic feedback for the integration of intentions in shared task execution. In: *IEEE transactions on haptics* 6 (2013), Nr. 1, S. 94–105. `http://dx.doi.org/10.1109/TOH.2012.2`. – DOI 10.1109/TOH.2012.2. – ISSN 1939–1412

[GHW+11] Geyer, S.; Hakuli, S.; Winner, H.; Franz, B.; Kauer, M.: Development of a cooperative system behavior for a highly automated vehicle guidance concept based on the Conduct-by-Wire principle. In: *2011 IEEE Intelligent Vehicles Symposium (IV)*, IEEE, 2011. – ISBN 978–1–4577–0890–9, S. 411–416

[GR82] Greenstein, J. S.; Rouse, W. B.: A Model of Human Decisionmaking in Multiple Process Monitoring Situations. In: *IEEE Transactions on Systems, Man, and Cybernetics* 12 (1982), Nr. 2, S. 182–193. `http://dx.doi.org/10.1109/TSMC.1982.4308802`. – DOI 10.1109/TSMC.1982.4308802. – ISSN 0018–9472

[GR86] Greenstein, J.; Revesman, M.: Development and Validation of a Mathematical Model of Human Decisionmaking for Human-Computer Communication. In: *IEEE Transactions on Systems, Man, and Cybernetics* 16 (1986), Nr. 1, S. 148–154. `http://dx.doi.org/10.1109/TSMC.1986.289291`. – DOI 10.1109/TSMC.1986.289291. – ISSN 0018–9472

[Gro11] Groten, R. K.: *Haptic Human-Robot Collaboration: How to Learn from Human Dyads*. München, Technische Universität München, Dissertation, 2011

[GS83] GENTNER, D. (Hrsg.); STEVENS, A. L. (Hrsg.): *Mental Models*. N. J.: Lawrence Erlbaum Associates, Inc., 1983. – ISBN 0898592429

[GV19] GENG, B.; VARSHNEY, P. K.: On Decision Making In Human-Machine Networks. In: *2019 IEEE 16th International Conference on Mobile Ad Hoc and Sensor Systems (MASS)*, IEEE, 2019. – ISBN 978–1–7281–4601–0, S. 37–45

[Har06] HART, S. G.: Nasa-Task Load Index (NASA-TLX); 20 Years Later. In: *Proceedings of the Human Factors and Ergonomics Society Annual Meeting* 50 (2006), Nr. 9, S. 904–908. http://dx.doi.org/10.1177/154193120605000909. – DOI 10.1177/154193120605000909. – ISSN 1541–9312

[Har17] HARRIS, C. M.: Autonomous Vehicle Decision-Making: Should We Be Bio-inspired? Version: 2017. http://dx.doi.org/10.1007/978-3-319-64107-2{_}25. In: GAO, Y. (Hrsg.); FALLAH, S. (Hrsg.); JIN, Y. (Hrsg.); LEKAKOU, C. (Hrsg.): *Towards Autonomous Robotic Systems* Bd. 10454. Cham: Springer International Publishing, 2017. – DOI 10.1007/978–3–319–64107–2_25. – ISBN 978–3–319–64106–5, S. 315–324

[Hei58] HEIDER, F.: *The psychology of interpersonal relations*. Hoboken: John Wiley & Sons Inc, 1958. http://dx.doi.org/10.1037/10628-000. http://dx.doi.org/10.1037/10628-000

[HG09] HARDIN, B.; GOODRICH, M. A.: On using mixed-initiative control. In: SCHEUTZ, M. (Hrsg.); MICHAUD, F. (Hrsg.); HINDS, P. (Hrsg.); SCASSELLATI, B. (Hrsg.): *Proceedings of the 4th ACM/IEEE international conference on Human robot interaction - HRI '09*. New York, New York, USA: ACM Press, 2009. – ISBN 9781605584041, S. 165

[HHBR15] HORNAUER, S.; HAHN, A.; BLAICH, M.; REUTER, J.: Trajectory Planning with Negotiation for Maritime Collision Avoidance. In: *TransNav, the International Journal on Marine Navigation and Safety of Sea Transportation* 9 (2015), Nr. 3, S. 335–341. http://dx.doi.org/10.12716/1001.09.03.05. – DOI 10.12716/1001.09.03.05. – ISSN 2083–6473

[HIL+19] HONG, A.; IGHARORO, O.; LIU, Y.; NIROUI, F.; NEJAT, G.; BENHABIB, B.: Investigating Human-Robot Teams for Learning-Based Semi-autonomous Control in Urban Search and Rescue Environments. In: *Journal of Intelligent & Robotic Systems* 94 (2019), Nr. 3-4, S. 669–686. http://dx.doi.org/10.1007/s10846-018-0899-0. – DOI 10.1007/s10846–018–0899–0. – ISSN 0921–0296

[HJ09] HINDRIKS, K. V.; JONKER, C. M.: Creating human-machine synergy in negotiation support systems. In: HINDRIKS, K. V. (Hrsg.); BRINKMAN,

W.-P. (Hrsg.): *Proceedings of the 1st International Working Conference on Human Factors and Computational Models in Negotiation - HuCom '08*. New York, New York, USA: ACM Press, 2009. – ISBN 9789081381116, S. 47–54

[HL14] HAO, J.; LEUNG, H.-f.: CUHKAgent: An Adaptive Negotiation Strategy for Bilateral Negotiations over Multiple Items. Version: 2014. `http://dx.doi.org/10.1007/978-4-431-54758-7{_}11`. In: MARSA-MAESTRE, I. (Hrsg.); LOPEZ-CARMONA, M. A. (Hrsg.); ITO, T. (Hrsg.); ZHANG, M. (Hrsg.); BAI, Q. (Hrsg.); FUJITA, K. (Hrsg.): *Novel Insights in Agent-based Complex Automated Negotiation* Bd. 535. Tokyo: Springer Japan, 2014. – DOI 10.1007/978–4–431–54758–7_11. – ISBN 978–4–431–54757–0, S. 171–179

[Hoc01] HOC, J.-M.: Towards a cognitive approach to human–machine cooperation in dynamic situations. In: *International Journal of Human-Computer Studies* 54 (2001), Nr. 4, S. 509–540. `http://dx.doi.org/10.1006/ijhc.2000.0454`. – DOI 10.1006/ijhc.2000.0454. – ISSN 10715819

[HS11] HÖRNER, J.; SAHUGUET, N.: A war of attrition with endogenous effort levels. In: *Economic Theory* 47 (2011), Nr. 1, S. 1–27. `http://dx.doi.org/10.1007/S00199-010-0524-3`. – DOI 10.1007/S00199–010–0524–3. – ISSN 0938–2259

[HSW05] HIMOFF, J.; SKOBELEV, P.; WOOLDRIDGE, M.: MAGENTA technology. In: PECHOUCEK, M. (Hrsg.); STEINER, D. (Hrsg.); THOMPSON, S. (Hrsg.): *Proceedings of the fourth international joint conference on Autonomous agents and multiagent systems - AAMAS '05*. New York, New York, USA: ACM Press, 2005. – ISBN 1595930930, S. 60

[HT08] HINDRIKS, K.; TYKHONOV, D.: Opponent Modelling in Automated Multi-issue Negotiation Using Bayesian Learning. In: *Proceedings of the 7th International Joint Conference on Autonomous Agents and Multiagent Systems - Volume 1*. Richland, SC: International Foundation for Autonomous Agents and Multiagent Systems, 2008 (AAMAS '08). – ISBN 978–0–9817381–0–9, 331–338

[IFH19] INGA, J.; FLAD, M.; HOHMANN, S.: Validation of a Human Cooperative Steering Behavior Model Based on Differential Games. In: *2019 IEEE International Conference on Systems, Man and Cybernetics (SMC)*, IEEE, 2019. – ISBN 978–1–7281–4569–3, S. 3124–3129

[Ing21] INGA CHARAJA, J. J.: *Inverse Dynamic Game Methods for Identification of Cooperative System Behavior*, Diss., 2021. `http://dx.doi.org/10.5445/KSP/1000128612`. – DOI 10.5445/KSP/1000128612

[JA15] Jiang, S.; Arkin, R. C.: Mixed-Initiative Human-Robot Interaction: Definition, Taxonomy, and Survey. In: *2015 IEEE International Conference on Systems, Man, and Cybernetics*, IEEE, 2015. – ISBN 978–1–4799–8697–2, S. 954–961

[JA19] Jain, S.; Argall, B.: Probabilistic Human Intent Recognition for Shared Autonomy in Assistive Robotics. In: *ACM transactions on human-robot interaction* 9 (2019), Nr. 1. http://dx.doi.org/10.1145/3359614. – DOI 10.1145/3359614

[JCH16] Johnson, L. B.; Choi, H.-L.; How, J. P.: The role of information assumptions in decentralized task allocation: A tutorial. In: *IEEE Control Systems* 36 (2016), Nr. 4, S. 45–58. http://dx.doi.org/10.1109/MCS.2016.2558419. – DOI 10.1109/MCS.2016.2558419. – ISSN 1066–033X

[JMS+16] Johns, M.; Mok, B.; Sirkin, D.; Gowda, N.; Smith, C.; Talamonti, W.; Ju, W.: Exploring shared control in automated driving. In: *2016 11th ACM/IEEE International Conference on Human-Robot Interaction (HRI)*, IEEE, 2016. – ISBN 978–1–4673–8370–7, S. 91–98

[JSB13] Jarrassé, N.; Sanguineti, V.; Burdet, E.: Slaves no longer: Review on role assignment for human–robot joint motor action. In: *Adaptive Behavior* 22 (2013), Nr. 1, S. 70–82. http://dx.doi.org/10.1177/1059712313481044. – DOI 10.1177/1059712313481044. – ISSN 1059–7123

[JWBA16] Jonge, A. W.; Wildenbeest, J. G. W.; Boessenkool, H.; Abbink, D. A.: The Effect of Trial-by-Trial Adaptation on Conflicts in Haptic Shared Control for Free-Air Teleoperation Tasks. In: *IEEE transactions on haptics* 9 (2016), Nr. 1, S. 111–120. http://dx.doi.org/10.1109/TOH.2015.2477302. – DOI 10.1109/TOH.2015.2477302. – ISSN 1939–1412

[KAL07] Krainin, M.; An, B.; Lesser, V.: An Application of Automated Negotiation to Distributed Task Allocation. In: *2007 IEEE/WIC/ACM International Conference on Intelligent Agent Technology (IAT'07)*, IEEE, 2007. – ISBN 0–7695–3027–3, S. 138–145

[KFS+12] Kauer, M.; Franz, B.; Schreiber, M.; Bruder, R.; Geyer, S.: User acceptance of cooperative maneuver-based driving–a summary of three studies. In: *Work (Reading, Mass.)* 41 Suppl 1 (2012), S. 4258–4264. http://dx.doi.org/10.3233/WOR-2012-0720-4258. – DOI 10.3233/WOR–2012–0720–4258

[Khe11] Kheddar, A.: Human-robot haptic joint actions is an equal control-sharing approach possible? In: *2011 4th International Conference on*

Human System Interactions, HSI 2011, IEEE, 2011. – ISBN 978–1–4244–9638–9, S. 268–273

[KM97] Krishna, V.; Morgan, J.: An Analysis of the War of Attrition and the All-Pay Auction. In: *Journal of Economic Theory* 72 (1997), Nr. 2, S. 343–362. http://dx.doi.org/10.1006/jeth.1996.2208. – DOI 10.1006/jeth.1996.2208. – ISSN 00220531

[KR14] Keeney, R. L.; Raiffa, H.: *Decisions with Multiple Objectives.* Cambridge University Press, 2014. http://dx.doi.org/10.1017/CBO9781139174084. http://dx.doi.org/10.1017/CBO9781139174084. – ISBN 9780521438834

[KSB10] Kauer, M.; Schreiber, M.; Bruder, R.: How to conduct a car? A design example for maneuver based driver-vehicle interaction. In: *2010 IEEE Intelligent Vehicles Symposium*, IEEE, 2010. – ISBN 978–1–4244–7866–8, S. 1214–1221

[KSB13] Kucukyilmaz, A.; Sezgin, T. M.; Basdogan, C.: Intention recognition for dynamic role exchange in haptic collaboration. In: *IEEE transactions on haptics* 6 (2013), Nr. 1, S. 58–68. http://dx.doi.org/10.1109/ToH.2012.21. – DOI 10.1109/ToH.2012.21. – ISSN 1939–1412

[KT79] Kahneman, D.; Tversky, A.: Prospect Theory: An Analysis of Decision under Risk. In: *Econometrica* 47 (1979), Nr. 2, S. 263. http://dx.doi.org/10.2307/1914185. – DOI 10.2307/1914185. – ISSN 00129682

[KW52] Kruskal, W. H.; Wallis, W. A.: Use of Ranks in One-Criterion Variance Analysis. In: *Journal of the American Statistical Association* 47 (1952), Nr. 260, S. 583. http://dx.doi.org/10.2307/2280779. – DOI 10.2307/2280779. – ISSN 01621459

[KW89] Kennan, J.; Wilson, R.: Strategic Bargaining Models and Interpretation of Strike Data. In: *Journal of Applied Econometrics* 4 (1989), S87-S130. http://www.jstor.org/stable/2096595. – ISSN 08837252

[Lam13] Lamb, F.: *Industrial automation: hands-on.* McGraw-Hill Education, 2013

[Lan02] Landau, K.: Usability criteria for intelligent driver assistance systems. In: *Theoretical Issues in Ergonomics Science* 3 (2002), Nr. 4, S. 330–345. http://dx.doi.org/10.1080/14639220210137748. – DOI 10.1080/14639220210137748. – ISSN 1463–922X

[LC10] Lopes, F.; Coelho, H.: Bilateral Negotiation in a Multi-agent Supply Chain System. Version: 2010. http://dx.doi.org/10.1007/

978-3-642-15208-5{_}18. In: van der Aalst, W. (Hrsg.); Mylopoulos, J. (Hrsg.); Sadeh, N. M. (Hrsg.); Shaw, M. J. (Hrsg.); Szyperski, C. (Hrsg.); Buccafurri, F. (Hrsg.); Semeraro, G. (Hrsg.): *E-Commerce and Web Technologies* Bd. 61. Berlin, Heidelberg: Springer Berlin Heidelberg, 2010. – DOI 10.1007/978–3–642–15208–5_18. – ISBN 978–3–642–15207–8, S. 195–206

[LHFH18] Ludwig, J.; Haas, A.; Flad, M.; Hohmann, S.: A Comparison of Concepts for Control Transitions from Automation to Human. In: *2018 IEEE International Conference on Systems, Man, and Cybernetics (SMC)*, IEEE, 2018. – ISBN 978–1–5386–6650–0, S. 3201–3206

[Lik32] Likert, R.: *A technique for the measurement of attitudes*. New York, NY, New York University, Dissertation, 1932

[May74] Maynard Smith, J.: The theory of games and the evolution of animal conflicts. In: *Journal of Theoretical Biology* 47 (1974), Nr. 1, S. 209–221. http://dx.doi.org/10.1016/0022-5193(74)90110-6. – DOI 10.1016/0022–5193(74)90110–6. – ISSN 00225193

[MCAV19] Morais, D. C.; Carreras, A.; Almeida, A. T.; Vetschera, R.: *Group Decision and Negotiation: Behavior, Models, and Support*. Bd. 351. Cham: Springer International Publishing, 2019. http://dx.doi.org/10.1007/978-3-030-21711-2. http://dx.doi.org/10.1007/978-3-030-21711-2. – ISBN 978–3–030–21710–5

[MEA92] Madden, T. J.; Ellen, P. S.; Ajzen, I.: A Comparison of the Theory of Planned Behavior and the Theory of Reasoned Action. In: *Personality and Social Psychology Bulletin* 18 (1992), Nr. 1, S. 3–9. http://dx.doi.org/10.1177/0146167292181001. – DOI 10.1177/0146167292181001. – ISSN 0146–1672

[MG17] Mell, J.; Gratch, J.: Grumpy & Pinocchio: Answering Human-Agent Negotiation Questions Through Realistic Agent Design. In: *Proceedings of the 16th Conference on Autonomous Agents and MultiAgent Systems*. Richland, SC: International Foundation for Autonomous Agents and Multiagent Systems, 2017 (AAMAS '17), 401–409

[MHM11] Masvoula, M.; Halatsis, C.; Martakos, D.: Predictive Automated Negotiators Employing Risk-Seeking and Risk-Averse Strategies. Version: 2011. http://dx.doi.org/10.1007/978-3-642-23957-1{_}37. In: Iliadis, L. (Hrsg.); Jayne, C. (Hrsg.): *Engineering Applications of Neural Networks* Bd. 363. Berlin, Heidelberg: Springer Berlin Heidelberg, 2011. – DOI 10.1007/978–3–642–23957–1_37. – ISBN 978–3–642–23956–4, S. 325–334

[Mic86] Michon, J. A.: A Critical View of Driver Behavior Models: What Do We Know, What Should We Do? Version: 1986. http://dx.doi.org/10.1007/978-1-4613-2173-6{_}19. In: Evans, L. (Hrsg.); Schwing, R. C. (Hrsg.): *Human Behavior and Traffic Safety*. Boston, MA: Springer US, 1986. – DOI 10.1007/978–1–4613–2173–6_19. – ISBN 978–1–4612–9280–7, S. 485–524

[MLG20] Mell, J.; Lucas, G. M.; Gratch, J.: Varied Magnitude Favor Exchange in Human-Agent Negotiation. In: *Proceedings of the 20th ACM International Conference on Intelligent Virtual Agents*. New York, NY, USA: ACM, 2020. – ISBN 9781450375863, S. 1–8

[MLH15] Medina, J. R.; Lorenz, T.; Hirche, S.: Synthesizing Anticipatory Haptic Assistance Considering Human Behavior Uncertainty. In: *IEEE Transactions on Robotics* 31 (2015), Nr. 1, S. 180–190. http://dx.doi.org/10.1109/TRO.2014.2387571. – DOI 10.1109/TRO.2014.2387571. – ISSN 1552–3098

[MLK+12] Mortl, A.; Lawitzky, M.; Kucukyilmaz, A.; Sezgin, M.; Basdogan, C.; Hirche, S.: The role of roles: Physical cooperation between humans and robots. In: *The International Journal of Robotics Research* 31 (2012), Nr. 13, S. 1656–1674. http://dx.doi.org/10.1177/0278364912455366. – DOI 10.1177/0278364912455366. – ISSN 0278–3649

[MLM+11] Medina, J. R.; Lawitzky, M.; Mortl, A.; Lee, D.; Hirche, S.: An experience-driven robotic assistant acquiring human knowledge to improve haptic cooperation. In: *2011 IEEE/RSJ International Conference on Intelligent Robots and Systems (IROS 2011)*, 2011, S. 2416–2422

[MM95] Millot, P.; Mandiau, R.: Man-machine cooperative organizations: Formal and pragmatic implementation methods. In: Hoc, J.-M. (Hrsg.); Cacciabue, P. C. (Hrsg.); Hollnagel, E. (Hrsg.): *Expertise and Technology*. Hillsdale, N.J.: Lawrence Erlbaum Associates, Inc., 1995, S. 213–228

[MO04] Marayong, P.; Okamura, A. M.: Speed-accuracy characteristics of human-machine cooperative manipulation using virtual fixtures with variable admittance. In: *Human factors* 46 (2004), Nr. 3, S. 518–532. http://dx.doi.org/10.1518/hfes.46.3.518.50400. – DOI 10.1518/hfes.46.3.518.50400. – ISSN 0018–7208

[Moo85] Moorthy, K. S.: Using Game Theory to Model Competition. In: *Journal of Marketing Research* 22 (1985), Nr. 3, S. 262. http://dx.doi.org/10.2307/3151424. – DOI 10.2307/3151424. – ISSN 00222437

[MP95] McKelvey, R. D.; Palfrey, T. R.: Quantal Response Equilibria for Normal Form Games. In: *Games and Economic Behavior* 10 (1995), Nr. 1, S. 6–38. http://dx.doi.org/10.1006/game.1995.1023. – DOI 10.1006/game.1995.1023. – ISSN 08998256

[MTD10] Mercier, S.; Tessier, C.; Dehais, F.: Authority management in human-robot systems. In: *IFAC Proceedings Volumes* 43 (2010), Nr. 13, S. 410–415. http://dx.doi.org/10.3182/20100831-4-FR-2021.00072. – DOI 10.3182/20100831–4–FR–2021.00072. – ISSN 14746670

[Mye91] Myerson, R. B.: *Game theory: Analysis of conflict*. Cambridge, Mass: Harvard University Press, 1991 https://search.ebscohost.com/login.aspx?direct=true&scope=site&db=nlebk&db=nlabk&AN=562543. – ISBN 9780674728615

[Nag95] Nagel, R.: Unraveling in Guessing Games: An Experimental Study. In: *The American Economic Review* 85 (1995), Nr. 5, 1313–1326. http://www.jstor.org/stable/2950991. – ISSN 00028282

[NC15] Na, X.; Cole, D. J.: Game-Theoretic Modeling of the Steering Interaction Between a Human Driver and a Vehicle Collision Avoidance Controller. In: *IEEE Transactions on Human-Machine Systems* 45 (2015), Nr. 1, S. 25–38. http://dx.doi.org/10.1109/THMS.2014.2363124. – DOI 10.1109/THMS.2014.2363124. – ISSN 2168–2291

[NFA08] Nielsen, C. W.; Few, D. A.; Athey, D. S.: Using mixed-initiative human-robot interaction to bound performance in a search task. In: *2008 International Conference on Intelligent Sensors, Sensor Networks and Information Processing*, IEEE, 2008. – ISBN 978–1–4244–2956–1, S. 195–200

[Nor83] Norman, D. A.: Design rules based on analyses of human error. In: *Communications of the ACM* 26 (1983), Nr. 4, S. 254–258. http://dx.doi.org/10.1145/2163.358092. – DOI 10.1145/2163.358092. – ISSN 0001–0782

[NSWS20] Nagahama, A.; Saito, T.; Wada, T.; Sonoda, K.: Autonomous Driving Learning Preference of Collision Avoidance Maneuvers. In: *IEEE Transactions on Intelligent Transportation Systems* (2020), S. 1–11. http://dx.doi.org/10.1109/TITS.2020.2988303. – DOI 10.1109/TITS.2020.2988303. – ISSN 1524–9050

[OGD17] Owan, P.; Garbini, J.; Devasia, S.: Addressing agent disagreement in mixed-initiative traded control for confined-space manufacturing. In: *2017 IEEE International Conference on Advanced Intelligent Mechatronics (AIM)*, IEEE, 2017. – ISBN 978–1–5090–5998–0, S. 227–234

[OKSB10] Oguz, S. O.; Kucukyilmaz, A.; Sezgin, T. M.; Basdogan, C.: Haptic negotiation and role exchange for collaboration in virtual environments. In: *2010 IEEE Haptics Symposium (Formerly known as Symposium on Haptic Interfaces for Virtual Environment and Teleoperator Systems)*, 2010, S. 371–378

[OKSB12] Oguz, S. O.; Kucukyilmaz, A.; Sezgin, T. M.; Basdogan, C.: Supporting Negotiation Behavior with Haptics-Enabled Human-Computer Interfaces. In: *IEEE transactions on haptics* 5 (2012), Nr. 3, S. 274–284. http://dx.doi.org/10.1109/TOH.2012.37. – DOI 10.1109/TOH.2012.37. – ISSN 1939–1412

[OLK09] Oshrat, Y.; Lin, R.; Kraus, S.: Facing the Challenge of Human-Agent Negotiations via Effective General Opponent Modeling. In: *Proceedings of The 8th International Conference on Autonomous Agents and Multiagent Systems - Volume 1*. Richland, SC: International Foundation for Autonomous Agents and Multiagent Systems, 2009 (AAMAS '09). – ISBN 9780981738161, S. 377–384

[PLD02] Pacaux-Lemoine, M. P.; Debernard, S.: Common work space for human–machine cooperation in air traffic control. In: *Control Engineering Practice* 10 (2002), Nr. 5, S. 571–576. http://dx.doi.org/10.1016/S0967-0661(01)00060-0. – DOI 10.1016/S0967–0661(01)00060–0. – ISSN 09670661

[PLF16] Pacaux-Lemoine, M.-P.; Flemisch, F.: Layers of Shared and Cooperative Control, assistance and automation. In: *IFAC-PapersOnLine* 49 (2016), Nr. 19, S. 159–164. http://dx.doi.org/10.1016/j.ifacol.2016.10.479. – DOI 10.1016/j.ifacol.2016.10.479. – ISSN 24058963

[PLHSC20] Pacaux-Lemoine, M.-P.; Habib, L.; Sciacca, N.; Carlson, T.: Emulated haptic shared control for brain-computer interfaces improves human-robot cooperation. In: *2020 IEEE International Conference on Human-Machine Systems (ICHMS)*, IEEE, 2020. – ISBN 978–1–7281–5871–6, S. 1–6

[PLI15] Pacaux-Lemoine, M.-P.; Itoh, M.: Towards Vertical and Horizontal Extension of Shared Control Concept. In: *2015 IEEE International Conference on Systems, Man, and Cybernetics*, IEEE, 2015. – ISBN 978–1–4799–8697–2, S. 3086–3091

[PLV13] Pacaux-Lemoine, M.-P.; Vanderhaegen, F.: Towards Levels of Cooperation. In: *2013 IEEE International Conference on Systems, Man, and Cybernetics*, IEEE, 2013. – ISBN 978–1–4799–0652–9, S. 291–296

[PSW00] Parasuraman, R.; Sheridan, T.; Wickens, C.: A model for types and levels of human interaction with automation. IEEE Trans. Syst.

Man Cybern. Part A Syst. Hum. 30(3), 286-297. In: *IEEE transactions on systems, man, and cybernetics. Part A, Systems and humans : a publication of the IEEE Systems, Man, and Cybernetics Society* 30 (2000), S. 286–297. `http://dx.doi.org/10.1109/3468.844354`. – DOI 10.1109/3468.844354

[Ras83] Rasmussen, J.: Skills, rules, and knowledge; signals, signs, and symbols, and other distinctions in human performance models. In: *IEEE Transactions on Systems, Man, and Cybernetics* SMC-13 (1983), Nr. 3, S. 257–266. `http://dx.doi.org/10.1109/TSMC.1983.6313160`. – DOI 10.1109/TSMC.1983.6313160. – ISSN 0018–9472

[RHS11] Rosen, J. (Hrsg.); Hannaford, B. (Hrsg.); Satava, R. M. (Hrsg.): *Surgical Robotics*. Boston, MA: Springer US, 2011. `http://dx.doi.org/10.1007/978-1-4419-1126-1`. `http://dx.doi.org/10.1007/978-1-4419-1126-1`. – ISBN 978–1–4419–1125–4

[Rin14] Rinne, H.: *The Hazard rate : Theory and inference (with supplementary MATLAB-Programs)*. Justus-Liebig-Universität, 2014

[RP08] Reed, K. B.; Peshkin, M. A.: Physical Collaboration of Human-Human and Human-Robot Teams. In: *IEEE transactions on haptics* 1 (2008), Nr. 2, S. 108–120. `http://dx.doi.org/10.1109/TOH.2008.13`. – DOI 10.1109/TOH.2008.13. – ISSN 1939–1412

[RS06] Ros, R.; Sierra, C.: A Negotiation Meta Strategy Combining Trade-off and Concession Moves. In: *Autonomous Agents and Multi-Agent Systems* 12 (2006), Nr. 2, S. 163–181. `http://dx.doi.org/10.1007/s10458-006-5837-z`. – DOI 10.1007/s10458–006–5837–z. – ISSN 1387–2532

[Rub82] Rubinstein, A.: Perfect Equilibrium in a Bargaining Model. In: *Econometrica* 50 (1982), Nr. 1, S. 97. `http://dx.doi.org/10.2307/1912531`. – DOI 10.2307/1912531. – ISSN 00129682

[Sar83] Saridis, G.: Intelligent robotic control. In: *IEEE Transactions on Automatic Control* 28 (1983), Nr. 5, S. 547–557. `http://dx.doi.org/10.1109/TAC.1983.1103278`. – DOI 10.1109/TAC.1983.1103278. – ISSN 0018–9286

[SBK06] Sebanz, N.; Bekkering, H.; Knoblich, G.: Joint action: bodies and minds moving together. In: *Trends in cognitive sciences* 10 (2006), Nr. 2, S. 70–76. `http://dx.doi.org/10.1016/j.tics.2005.12.009`. – DOI 10.1016/j.tics.2005.12.009. – ISSN 1364–6613

[SBP+18] Scholtens, W.; Barendswaard, S.; Pool, D.; van Paassen, R.; Abbink, D.: A New Haptic Shared Controller Reducing Steering Conflicts. In: *2018 IEEE International Conference on Systems, Man, and Cybernetics (SMC)*, IEEE, 2018. – ISBN 978–1–5386–6650–0, S. 2705–2710

[SGC98] Stuhlmacher, A. F.; Gillespie, T. L.; Champagne, M. V.: The impact of time pressure in negotiation: a meta-analysis. In: *International Journal of Conflict Management* 9 (1998), Nr. 2, S. 97–116. http://dx.doi.org/10.1108/eb022805. – DOI 10.1108/eb022805

[She11] Sheridan, T. B.: Adaptive Automation, Level of Automation, Allocation Authority, Supervisory Control, and Adaptive Control: Distinctions and Modes of Adaptation. In: *IEEE Transactions on Systems, Man, and Cybernetics - Part A: Systems and Humans* 41 (2011), Nr. 4, S. 662–667. http://dx.doi.org/10.1109/TSMCA.2010.2093888. – DOI 10.1109/TSMCA.2010.2093888. – ISSN 1083–4427

[SLB09] Shoham, Y.; Leyton-Brown, K.: *Multiagent systems: Algorithmic, game-theoretic, and logical foundations*. Cambridge: Cambridge University Press, 2009 https://doi.org/10.1017/CBO9780511811654. – ISBN 0521899435

[SLL78] Sheridan, T.; L. Verplank, W.; L. Brooks, T.: Human and Computer Control of Undersea Teleoperators. (1978)

[SRBL91] Schmidt, K.; Rasmussen, J.; Brehmer, B.; Leplat, J.: Cooperative work: A conceptual framework. In: *Distributed decision making: Cognitive models for cooperative work* (1991), S. 75–110

[Str14] Strzalecki, T.: Depth of reasoning and higher order beliefs. In: *Journal of Economic Behavior & Organization* 108 (2014), S. 108–122. http://dx.doi.org/10.1016/j.jebo.2014.09.002. – DOI 10.1016/j.jebo.2014.09.002. – ISSN 01672681

[SVP11] Sislak, D.; Volf, P.; Pechoucek, M.: Agent-Based Cooperative Decentralized Airplane-Collision Avoidance. In: *IEEE Transactions on Intelligent Transportation Systems* 12 (2011), Nr. 1, S. 36–46. http://dx.doi.org/10.1109/TITS.2010.2057246. – DOI 10.1109/TITS.2010.2057246. – ISSN 1524–9050

[SWS19] Suehiro, Y.; Wada, T.; Sonoda, K.: Assistance Method for Merging by Increasing Clarity of Decision Making. In: *IEEE Transactions on Intelligent Vehicles* 4 (2019), Nr. 1, S. 48–58. http://dx.doi.org/10.1109/TIV.2018.2886692. – DOI 10.1109/TIV.2018.2886692. – ISSN 2379–8904

[TGS11] THOBBI, A.; GU, Y.; SHENG, W.: Using human motion estimation for human-robot cooperative manipulation. In: *2011 IEEE/RSJ International Conference on Intelligent Robots and Systems*, IEEE, 2011. – ISBN 978–1–61284–456–5, S. 2873–2878

[TI17] TANG, F.; ITO, E.: Human-assisted navigation through sliding autonomy. In: *2017 2nd International Conference on Robotics and Automation Engineering (ICRAE)*, IEEE, 2017. – ISBN 978–1–5386–1306–1, S. 26–30

[TLL+18] TIAN, R.; LI, S.; LI, N.; KOLMANOVSKY, I.; GIRARD, A.; YILDIZ, Y.: Adaptive Game-Theoretic Decision Making for Autonomous Vehicle Control at Roundabouts. In: *2018 IEEE Conference on Decision and Control (CDC)*, IEEE, 2018. – ISBN 978–1–5386–1395–5, S. 321–326

[Tuk97] TUKEY, J. W.: *Exploratory data analysis*. Reading, Mass.: Addison-Wesley, 1997 (Addison-Wesley series in behavioral science Quantitative methods). – ISBN 0201076160

[TW19] TURNWALD, A.; WOLLHERR, D.: Human-Like Motion Planning Based on Game Theoretic Decision Making. In: *International Journal of Social Robotics* 11 (2019), Nr. 1, S. 151–170. `http://dx.doi.org/10.1007/s12369-018-0487-2`. – DOI 10.1007/s12369–018–0487–2. – ISSN 1875–4791

[van12] VAN DER VAART, A. W.: *Asymptotic Statistics*. Cambridge University Press, 2012. `http://dx.doi.org/10.1017/CBO9780511802256`. `http://dx.doi.org/10.1017/CBO9780511802256`. – ISBN 9780521496032

[VGLH11] VALERO-GOMEZ, A.; LA PUENTE, P. de; HERNANDO, M.: Impact of two adjustable-autonomy models on the scalability of single-human/multiple-robot teams for exploration missions. In: *Human factors* 53 (2011), Nr. 6, S. 703–716. `http://dx.doi.org/10.1177/0018720811420427`. – DOI 10.1177/0018720811420427. – ISSN 0018–7208

[VKG14] VAHIDOV, R.; KERSTEN, G.; GUIMON, D.: Bilateral and Multi-bilateral Agent-Human Negotiations: Two Experiments. In: *2014 47th Hawaii International Conference on System Sciences*, IEEE, 2014. – ISBN 978–1–4799–2504–9, S. 241–249

[VNE+01] VOLPE, R.; NESNAS, I.; ESTLIN, T.; MUTZ, D.; PETRAS, R.; DAS, H.: The CLARAty architecture for robotic autonomy. In: *2001 IEEE Aerospace Conference Proceedings (Cat. No.01TH8542)*, IEEE, 2001. – ISBN 0–7803–6599–2, S. 1/121–1/132

[WAS+19] WESSEL, G.; ALTENDORF, E.; SCHWALM, M.; CANPOLAT, Y.; BURGHARDT, C.; FLEMISCH, F.: Self-determined nudging: a system concept for human–machine interaction. In: *Cognition, Technology &*

Work 21 (2019), Nr. 4, S. 621–630. `http://dx.doi.org/10.1007/s10111-019-00556-5`. – DOI 10.1007/s10111–019–00556–5. – ISSN 1435–5566

[WCW19] Walch, M.; Colley, M.; Weber, M.: Driving-task-related human-machine interaction in automated driving. In: Janssen, C. P. (Hrsg.); Donker, S. F. (Hrsg.); Chuang, L. L. (Hrsg.); Ju, W. (Hrsg.): *Proceedings of the 11th International Conference on Automotive User Interfaces and Interactive Vehicular Applications Adjunct Proceedings - AutomotiveUI '19*. New York, New York, USA: ACM Press, 2019. – ISBN 9781450369206, S. 427–433

[WHLS16] Winner, H. (Hrsg.); Hakuli, S. (Hrsg.); Lotz, F. (Hrsg.); Singer, C. (Hrsg.): *Handbook of Driver Assistance Systems: Basic Information, Components and Systems for Active Safety and Comfort*. Cham: Springer International Publishing, 2016. – ISBN 978–3–319–12352–3

[Wie61] Wiener, N.: *Cybernetics : or control and communication in the animal and the machine*. 2. ed. New York, NY: M.I.T. Press [u.a.], 1961

[WLCW19] Walch, M.; Lehr, D.; Colley, M.; Weber, M.: Don't you see them? In: Janssen, C. P. (Hrsg.); Donker, S. F. (Hrsg.); Chuang, L. L. (Hrsg.); Ju, W. (Hrsg.): *Proceedings of the 11th International Conference on Automotive User Interfaces and Interactive Vehicular Applications Adjunct Proceedings - AutomotiveUI '19*. New York, New York, USA: ACM Press, 2019. – ISBN 9781450369206, S. 232–237

[WSA+18] Wessel, G.; Schreck, C.; Altendorf, E.; Canpolat, Y.; Flemisch, F.: Learning from the Best – Naturalistic Arbitration for Cooperative Driving. In: Stanton, N. A. (Hrsg.): *Advances in Human Aspects of Transportation*. Cham: Springer International Publishing, 2018. – ISBN 978–3–319–60441–1, S. 631–642

[WSH+16] Walch, M.; Sieber, T.; Hock, P.; Baumann, M.; Weber, M.: Towards Cooperative Driving. In: Green, P. (Hrsg.); Boll, S. (Hrsg.); Burnett, G. (Hrsg.); Gabbard, J. (Hrsg.); Osswald, S. (Hrsg.): *Proceedings of the 8th International Conference on Automotive User Interfaces and Interactive Vehicular Applications - Automotive'UI 16*. New York, New York, USA: ACM Press, 2016. – ISBN 9781450345330, S. 261–268

[WVI04] Wollkind, S.; Valasek, J.; Ioerger, T.: Automated Conflict Resolution for Air Traffic Management Using Cooperative Multiagent Negotiation. In: *AIAA Guidance, Navigation, and Control Conference and Exhibit*. Reston, Virigina: American Institute of Aeronautics and Astronautics, 2004. – ISBN 978–1–62410–073–4

[WWM+19] Walch, M.; Woide, M.; Mühl, K.; Baumann, M.; Weber, M.: Cooperative Overtaking. In: *Proceedings of the 11th International Conference on Automotive User Interfaces and Interactive Vehicular Applications - AutomotiveUI '19*. New York, New York, USA: ACM Press, 2019. – ISBN 9781450368841, S. 144–155

[WWY11] Wang, G.; Wong, T. N.; Yu, C.: Computational method for agent-based E-commerce negotiations with adaptive negotiation behaviors. In: *Procedia Computer Science* 4 (2011), S. 1834–1843. `http://dx.doi.org/10.1016/j.procs.2011.04.199`. – DOI 10.1016/j.procs.2011.04.199. – ISSN 18770509

[YAB14] Yildiz, Y.; Agogino, A.; Brat, G.: Predicting Pilot Behavior in Medium-Scale Scenarios Using Game Theory and Reinforcement Learning. In: *Journal of Guidance, Control, and Dynamics* 37 (2014), Nr. 4, S. 1335–1343. `http://dx.doi.org/10.2514/1.G000176`. – DOI 10.2514/1.G000176. – ISSN 0731–5090

[YHS07] Yang, C.; Hu, Q.; Shi, C.: Automated Collision Resolution for Vessel Traffic Management by Using Cooperative Multi-Agent Negotiation. In: *2007 7th International Conference on ITS Telecommunications*, IEEE, 2007. – ISBN 1–4244–1177–7, S. 1–6

[Zeu19] Zeuthen, F.: *Problems of Monopoly and Economic Warfare*. [S.l.]: CRC PRESS, 2019. – ISBN 0815371586

[ZR89] Zlotkin, G.; Rosenschein, J. S.: Negotiation and Task Sharing Among Autonomous Agents in Cooperative Domains. In: *Proceedings of the 11th International Joint Conference on Artificial Intelligence - Volume 2*. San Francisco, CA, USA: Morgan Kaufmann Publishers Inc, 1989 (IJCAI'89), 912–917

Own Publications and Conference Contributions

[FRDH14] Flad, M.; Rothfuss, S.; Diehm, G.; Hohmann, S.: Active Brake Pedal Feedback Simulator Based on Electric Drive. In: *SAE International Journal of Passenger Cars - Electronic and Electrical Systems* 7 (2014), Nr. 1, S. 189–200

[KIR+17] Köpf, F.; Inga, J.; Rothfuss, S.; Flad, M.; Hohmann, S.: Inverse Reinforcement Learning for Identification in Linear-Quadratic Dynamic Games. In: *IFAC-PapersOnLine* 50 (2017), Nr. 1, S. 14902–14908

[MPMR16] Mayer, S.; Plangger, D.; Michahelles, F.; Rothfuss, S.: UberManufacturing: A goal-driven collaborative industrial manufacturing marketplace. In: Schmidt, A. (Hrsg.); Michahelles, F. (Hrsg.); Schneegass, S. (Hrsg.); Kritzler, M. (Hrsg.); Ilic, A. (Hrsg.); Kunze, K. (Hrsg.): *Proceedings of the 6th International Conference on the Internet of Things*. New York, NY, USA: ACM, 2016, S. 111–119

[RAFH20] Rothfuss, S.; Ayllon, C.; Flad, M.; Hohmann, S.: Adaptive Negotiation Model for Human-Machine Interaction on Decision Level. In: *2020 IFAC World Congress*, IFAC, 2020

[RCI+22] Rothfuss, S.; Chiou, M.; Inga, J.; Hohmann, S.; Stolkin, R.: A Negotiation-Theoretic Framework for Control Authority Transfer in Mixed-Initiative Robotic Systems. In: *2022 IEEE International Conference on Systems, Man, and Cybernetics (SMC)*. 2022

[RGFH18] Rothfuss, S.; Grauer, F.; Flad, M.; Hohmann, S.: A Steering Experiment Towards Haptic Cooperative Maneuver Negotiation. In: *2018 IEEE International Conference on Systems, Man, and Cybernetics (SMC)*, IEEE, 2018, S. 3207–3212

[RIK+17] Rothfuss, S.; Inga, J.; Köpf, F.; Flad, M.; Hohmann, S.: Inverse Optimal Control for Identification in Non-Cooperative Differential Games. In: *IFAC-PapersOnLine* 50 (2017), Nr. 1, S. 14909–14915

[RPFH19a] Rothfuss, S.; Prezdnyakov, R.; Flad, M.; Hohmann, S.: Decentralized path planning for cooperating autonomous mobile units. In: *Forschung im Ingenieurwesen* 83 (2019), Nr. 2, S. 137–147

[RPFH19b] Rothfuss, S.; Prezdnyakov, R.; Flad, M.; Hohmann, S.: Decentralized Path Planning for Cooperating Autonomous Vehicles. In: *AUTOREG, Mannheim, 2-3 Juli, 2019* Bd. 2349, VDI Verlag, 2019 (VDI-Berichte), S. 271–284

[RSFH19] Rothfuss, S.; Schmidt, R.; Flad, M.; Hohmann, S.: A Concept for Human-Machine Negotiation in Advanced Driving Assistance Systems. In: *2019 IEEE International Conference on Systems, Man and Cybernetics (SMC)*, IEEE, 2019, S. 3116–3123

[RSFH20] Rothfuss, S.; Steinkamp, J.; Flad, M.; Hohmann, S.: Cooperative Decision Making in Cooperative Control Systems by Means of Game Theory. In: Petrosyan, L. A. (Hrsg.); Mazalov, V. V. (Hrsg.); Zenkevich, N. A. (Hrsg.): *Frontiers of Dynamic Games*. Cham: Springer International Publishing, 2020 (Static & Dynamic Game Theory: Foundations & Applications), S. 245–265

[RTIH20] Rothfuss, S.; Tanaka, T. S.; Inga, J.; Hohmann, S.: The n-Stage War of Attrition and its Inverse Game Towards its Application in Human-Machine Cooperative Decision Making. In: *2020 59th IEEE Conference on Decision and Control (CDC)*, IEEE, 2020, S. 5386–5393

[RWI+] Rothfuss, S.; Wörner, M.; Inga, J.; Kiesel, A.; Hohmann, S.: Human-Machine Cooperative Decision Making Outperforms Individualism and Autonomy (under review). In: *IEEE Transactions on Human-Machine Systems*

[RWIH20] Rothfuss, S.; Wörner, M.; Inga, J.; Hohmann, S.: A Study on Human-Machine Cooperation on Decision Level. In: *2020 IEEE International Conference on Systems, Man, and Cybernetics (SMC)*, IEEE, 2020, S. 2291–2298

Supervised Theses

[AW19] AYLLON WRIEDT, C.: *An Adaptive Negotiation Concept for Human-Machine Interaction on Decision Level in the Context of Autonomous Driving*, Institute of Control Systems (IRS), Karlsruhe Institute of Technology (KIT), Master Thesis, 2019

[Bla19] BLANCO, A.: *Setup of an Experiment for Evaluating Cooperative Driving Assistance Systems*, Institute of Control Systems (IRS), Karlsruhe Institute of Technology (KIT), Bachelor Thesis, 2019

[Gra18] GRAUER, F.: *Design and Evaluation of an Experiment for Cooperative Vehicle Steering*, Institute of Control Systems (IRS), Karlsruhe Institute of Technology (KIT), Bachelor Thesis, 2018

[Ill17] ILLG, C.: *Development of Planning and Identifiction Methods for Trajectories*, Institute of Control Systems (IRS), Karlsruhe Institute of Technology (KIT), Bachelor Thesis, 2017

[Pre17] PREZDNYAKOV, R.: *Distributed Trajectory Planning for Cooperative Systems*, Institute of Control Systems (IRS), Karlsruhe Institute of Technology (KIT), Bachelor Thesis, 2017

[Sch18] SCHMIDT, R.: *Cooperative Driving Assistance System Development based on Negotiation Methods*, Institute of Control Systems (IRS), Karlsruhe Institute of Technology (KIT), Master Thesis, 2018

[Ste18] STEINKAMP, J.: *Negotiation Modeling by means of Game Theory for Cooperative Control Systems*, Institute of Control Systems (IRS), Karlsruhe Institute of Technology (KIT), Master Thesis, 2018

[Tan20] TANAKA, T. S.: *Design and Analysis of a Game-Theoretic Model Describing Cooperative Decision Making*, Institute of Control Systems (IRS), Karlsruhe Institute of Technology (KIT), Master Thesis, 2020

[Wör20] WÖRNER, M.: *Study of Human-Machine Cooperation on Decision Level Applied to Highly Automated Driving*, Institute of Control Systems (IRS), Karlsruhe Institute of Technology (KIT), Master Thesis, 2020

[Xue17] XUE, S.: *Development of Cooperative Coordination Concepts for Autonomous Taxis*, Institute of Control Systems (IRS), Karlsruhe Institute of Technology (KIT), Bachelor Thesis, 2017

Karlsruher Beiträge zur Regelungs- und Steuerungstechnik (ISSN 2511-6312)
Institut für Regelungs- und Steuerungssysteme

Band 01 Diehm, Gunter
Identifikation des menschlichen Bewegungsverhaltens auf der Basis von Primitiven.
ISBN 978-3-7315-0608-9

Band 02 Flad, Michael
Kooperative Regelungskonzepte auf Basis der Spieltheorie und deren Anwendung auf Fahrerassistenzsysteme.
ISBN 978-3-7315-0610-2

Band 03 Eckert, Marius
Modellbasierte Identifikation fraktionaler Systeme und ihre Anwendung auf die Lithium-Ionen-Zelle.
ISBN 978-3-7315-0690-4

Band 04 Krebs, Stefan
Intervallbeobachter für lineare parametervariante Systeme und deren Anwendung auf die Asynchronmaschine.
ISBN 978-3-7315-0857-1

Band 05 Kaspar, Stephan
Fahrdynamikuntersuchungen eines Elektrofahrzeugs mit Einzelrad-Hinterradantrieb.
ISBN 978-3-7315-0916-5

Band 06 Sauter, Patrick S.
Modellierung und zentrale prädiktive Regelung von multimodalen Energieverteilnetzen.
ISBN 978-3-7315-0963-9

Band 07 Kupper, Martin
Verteilte Zustandsschätzung fraktionaler Systeme und ihre Anwendung auf Lithium-Ionen-Batteriesysteme.
ISBN 978-3-7315-0971-4

Band 08 Merkert, Lennart
Optimal Scheduling of Combined Heat and Power Generation Considering Heating Grid Dynamics.
ISBN 978-3-7315-1056-7

Die Bände sind unter www.ksp.kit.edu als PDF frei verfügbar oder als Druckausgabe bestellbar.

Band 09 Ludwig, Julian
Automatisierte kooperative Transition einer Regelungsaufgabe zwischen Mensch und Maschine am Beispiel des hochautomatisierten Fahrens.
ISBN 978-3-7315-1069-7

Band 10 Inga Charaja, Juan Jairo
Inverse Dynamic Game Methods for Identification of Cooperative System Behavior.
ISBN 978-3-7315-1080-2

Band 11 Schnurr, Christoph Xaver
Ein Verfahren zur lexikographischen modellprädiktiven Regelung mit der Anwendung auf eine permanenterregte Synchronmaschine.
ISBN 978-3-7315-1095-6

Band 12 Schwab, Stefan
Guaranteed Verification of Dynamic Systems.
ISBN 978-3-7315-0965-3

Band 13 Pfeifer, Martin
Automated Model Generation and Observer Design for Interconnected Systems: A Port-Hamiltonian Approach.
ISBN 978-3-7315-1135-9

Band 14 König, Alexander
Absicherung hochautomatisierten Fahrens durch passiven virtuellen Dauerlauftest.
ISBN 978-3-7315-1141-0

Band 15 Stark, Oliver
Parameter- und Ordnungsidentifikation von fraktionalen Systemen mit einer Anwendung auf eine Lithium-Ionen-Batteriezelle.
ISBN 978-3-7315-1187-8

Band 16 Köpf, Florian
Adaptive Dynamic Programming: Solltrajektorienfolgeregelung und Konvergenzbedingungen.
ISBN 978-3-7315-1193-9

Band 17 Kölsch, Lukas
Dynamic Incentives for Optimal Control of Competitive Power Systems.
ISBN 978-3-7315-1209-7

Band 18 Schwartz, Manuel
Topologie-Optimierung eines radselektiv angesteuerten Fahrzeugs basierend auf einer optimalen Fahrzeugführungsregelung.
ISBN 978-3-7315-1222-6

Band 19 Rothfuß, Simon
Human-Machine Cooperative Decision Making.
ISBN 978-3-7315-1223-3
